AIR, WATER AND SOIL POLLUTION SCIENCE AND TECHNOLOGY

MARINE POLLUTION

TYPES, ENVIRONMENTAL SIGNIFICANCE AND MANAGEMENT STRATEGIES

AIR, WATER AND SOIL POLLUTION SCIENCE AND TECHNOLOGY

Additional books in this series can be found on Nova's website under the Series tab.

Additional e-books in this series can be found on Nova's website under the e-book tab.

AIR, WATER AND SOIL POLLUTION SCIENCE AND TECHNOLOGY

MARINE POLLUTION

TYPES, ENVIRONMENTAL SIGNIFICANCE AND MANAGEMENT STRATEGIES

DOMINIC E. JEFFERSON
EDITOR

nova publishers
New York

For permission to use material from this book please contact us:
Telephone 631-231-7269; Fax 631-231-8175
Web Site: http://www.novapublishers.com

Library of Congress Cataloging-in-Publication Data

ISBN: 978-1-63321-520-7

Published by Nova Science Publishers, Inc. † New York

CONTENTS

PREFACE

Despite the fact that water covers two-thirds of the Earth's surface, it is surprisingly vulnerable to human influence and activity. Marine ecosystems receive large amounts of a variety of pollutants from either treated or untreated wastewater. The extensive use of chemicals and their introduction into the environment has resulted in increased public concern for the potential threats to ecosystems, human health and safety. Persistent organic pollutants such as polychlorinated biphenyls, polycyclic aromatic hydrocarbons, polybrominated diphenyl ethers and pesticides, as well as heavy metals, surfactants; endocrine disrupting chemicals or personal care products and pharmaceuticals are pollutants usually present in marine environments. This book focuses on discussing the types of marine pollutants. It provides topics on the environmental significance and management strategies available.

Chapter 1 – Organotin compounds are the world's most widely used organometallic chemicals with annual production at approximately 50,000 tons. Pollution by organotins in the marine environment is generally associated with the historical use of tributyltin (TBT) and triphenyltin (TPhT) as antifoulants. Organotin compounds are ubiquitous within the coastal marine environment, with the highest concentration found within benthic sediments in regions of high boating activity and industry (ng g^{-1} to μg g^{-1} range). TBT and TPhT have strong affinities to particulate organic matter and demonstrate long-term persistence within the marine environment. Sediments are the major sink for TBT and TPhT, with rapid removal to the particulate matter phase from the water column generally preceding its degradation. Adsorption mechanisms are a dualistic function of hydrophobic interaction and electrostatic interaction. Despite the strong affinity to the solid-phase, adsorption is a reversible reaction, where desorption to the water column can readily occur. Due to its high toxicity, modern analytical techniques are required to be sensitive and selective enough to meet higher monitoring demands set by regulators, with gas chromatography and high performance liquid chromatography as the most commonly used techniques. UK management strategies for organotin compounds are incorporated within dredged material waste management. Work within this chapter aims to provide an overview of organotins within the marine environment, the monitoring and analytical methods used and the legislation governing dredging and the disposal of dredged material waste.

Chapter 2 – Large oil spills produce major ecological and economic impacts on shorelines depending on the following conditions: (i) persistence of the oil in the environment; (ii) weather and sea conditions; (iii) amount and rate of spillage; (iv) geographical, biological and economic characteristics of the area affected; (v) effectiveness of clean up. Oil spill dispersion models are necessary tools to forecast the trajectory and persistence of oil slicks following a marine accident. However, their reliability depends on the quality of available meteorological and hydrological data and assimilation strategies. Large oil spills that took place in regions relatively distant from the coast, such as the Prestige accident in November 2002 and the Deep Water Horizon explosion in April 2010, showed the need for having both a short-term forecast for the days following the accident as well as a long-term forecast to prevent future pollution impact. However, there is still no universally accepted methodology for the management of this kind of oil spill. A combined forecast of the short-term deterministic trajectory and the long-term probabilistic dispersion of the slicks may be useful to delimit the most probable consequences of any spill. Medess-4MS is an example of an integrated operational multi-model oil spill system which is under development to meet both requirements. This double methodology is illustrated by the analysis of the oil spill produced by the supertanker Prestige near the Spanish coast in November 2002. This double forecast makes it possible to obtain the short-term and long-term probability of oil beaching on different shoreline segments, followed by preliminary estimates of ecological risk, and economic cost based on the analysis of past similar accidents.

Chapter 3 – Despite the fact that water covers two-thirds of the Earth's surface, it is surprisingly vulnerable to human influence and activity. Marine ecosystems receive large amounts of a variety of pollutants from either treated or untreated wastewater. The extensive use of chemicals and their introduction into the environment has resulted in increased public concern for the potential threats to ecosystems, human health and safety. Persistent organic pollutants such as polychlorinated biphenyls, polycyclic aromatic hydrocarbons, polybrominated diphenyl ethers and pesticides, as well as heavy metals, surfactants; endocrine disrupting chemicals or personal care products and pharmaceuticals are pollutants usually present in marine environments. In recent years, as a result of this concern for environmental pollutants, there is a need to study and control the presence of a large amount of compounds in different environmental compartments, as many of them show carcinogenic, mutagenic or endocrine activity on living organisms. Wastewater discharges to the sea are the main route for these compounds to reach marine environments. In addition, the cumulative effects of these pollutants on coastal environments are expected to be considerable. Pollutants are always present in sediments or waters and from where they can reach animals and plants. Phyto and zooplankton absorb and accumulate toxic chemicals. Plankton is eaten by small fish that are, in turn, eaten by larger fish. As consequence, the toxic compound enters the food chain in high concentrations due to the effect of biological magnification. Over the last few years, legislation has limited the use of many chemicals because of their potential to produce adverse effects on wildlife and ecosystems. In this context, it is necessary the development and validation of analytical methods for the detection of these compounds in different marine compartments, which allows us to know the fate of these chemicals, in order to prevent, limit or reduce the impact of environmental contamination. The aim of the present work is to provide a comprehensive overview of the recent developments related to the determination of pollutants in the different marine compartments (water, sediments and biota), with special emphasis on sample preparation and analytical techniques as well as the achieved detection limits.

Chapter 4 – Oil related activities are potential threats for causing oil spills with serious environmental, economic and social impact. Oil spill pollution has been an issue of concern to coastal states as well as to international organizations and therefore work on prevention, detection, toxicity monitoring and impact assessment have been implemented at both national and international level. The contribution of new technologies, especially Geographical Information Systems and remote sensing, has improved both tracking and management operations. In the present work, oil behavior, monitoring and pollution impact of the marine environment have been reviewed. The importance of applying remote sensing and mapping techniques, modeling and GIS tools to either forecast oil trajectories or assess the vulnerability of coastal areas to oil and the impact on the marine ecosystem are presented. The advantages and shortcomings of different methodologies are discussed; however, the need for the implementation of integrated approaches seems the most effective approach. The advantage of applying such approaches in the framework of a GIS system and especially a webGIS system are discussed, since direct communication and interaction among authorities, stakeholders and the wider public is essential in the case of an emergency situation.

In: Marine Pollution
Editor: Dominic E. Jefferson
ISBN: 978-1-63321-520-7

Chapter 1

Organotin Compounds in the Marine Environment – Current Monitoring Techniques and Their Regulatory and Environmental Management Requirements in England and Wales

R. F. Cole[1], G. R. Fones[1], G. A. Mills[2], R. Parker[3], T. Bolam[3], A. C. Birchenough[3] and S. Kröger[3]

[1]School of Earth and Environmental Sciences,
University of Portsmouth, Portsmouth
[2]School of Pharmacy and Biomedical Sciences,
University of Portsmouth, Portsmouth
[3]Centre for Environment,
Fisheries and Aquaculture Science (Cefas),
Lowestoft Laboratory, Pakefield Road, Lowestoft, Suffolk, UK

Abstract

Organotin compounds are the world's most widely used organometallic chemicals with annual production at approximately 50,000 tons. Pollution by organotins in the marine environment is generally associated with the historical use of tributyltin (TBT) and triphenyltin (TPhT) as antifoulants. Organotin compounds are ubiquitous within the coastal marine environment, with the highest concentration found within benthic sediments in regions of high boating activity and industry (ng g^{-1} to μg g^{-1} range). TBT and TPhT have strong affinities to particulate organic matter and demonstrate long-term persistence within the marine environment. Sediments are the major sink for TBT and TPhT, with rapid removal to the particulate matter phase from the water column generally preceding its degradation. Adsorption mechanisms are a dualistic function of hydrophobic interaction and electrostatic interaction. Despite the strong affinity to the solid-phase, adsorption is a reversible reaction, where desorption to the water column can readily occur. Due to its high toxicity, modern analytical techniques are required to be sensitive and selective enough to meet higher monitoring demands set by regulators, with gas chromatography and high performance liquid chromatography as the most commonly used techniques. UK management strategies for organotin compounds are incorporated within dredged material waste management. Work within this chapter aims to provide an overview of organotins within the marine environment, the monitoring and analytical methods used and the legislation governing dredging and the disposal of dredged material waste.

1. INTRODUCTION

Historically organotin compounds have been mainly used in anti-fouling products to prevent fouling on ships. It has also been used in wood preservation, and the protection of textiles, paper and pulp (see Table 2). Organotin compounds have now been banned completely for use in anti-fouling paints under the International Convention on the Control of Harmful Anti-fouling Systems on Ships (AFS Convention Annex 1, 2001) (Anti-fouling systems, 2014) and European regulatory controls prevent their use in products for the EU market.

Whilst the use of organotins has now been restricted, it is highly persistent in the marine environment. Tri-substituted organotins, such as tributyltin (TBT) and triphenyltin (TPhT), have a long-term environmental persistence, particularly within anoxic sediments as they exhibit a strong affinity towards the solid phase. Therefore a legacy of significant concentrations of TBT remains in sediments in localised areas such as ports and harbours where ship building and repair and/or large volume vessel traffic is, or has been, prevalent (Bolam *et al* 2013). In certain areas where navigational dredging is required for vessel access, TBT concentrations remain high enough to deem the sediments too contaminated for disposal at sea. This can have significant economic implications, as alternatives to sea disposal (i.e., treating the material to reduce contamination the levels, disposal to land fill, capping at offshore disposal sites) are generally far more expensive (Bolam *et al* 2013).

Currently, there are no specific EU regulatory limits for organotins in marine sediments; however environmental quality standards (EQS) for surface waters exist within the EU Water Framework Directive (for TBT, a limit of 0.2 ng L^{-1} annual average concentration and 1.5 ng L^{-1} for maximum allowable concentrations). At a regional sea level, the Oslo and Paris commission (OSPAR) provide the legal mechanisms in guiding international cooperation for the protection of the marine environment in the North-East Atlantic (www.ospar.org). OSPAR supports EU legislation and incorporates it as part of the Coordinated Environmental Monitoring Programme (CEMP). OSPAR have issued Environmental Assessment Criteria (EACs) for TBT in different compartments. These criteria do not contribute to any legal mechanism; however provide a guideline for preliminary assessment of Joint Assessment and Monitoring Programme (JAMP) data. EACs for TBT (as cation) are given as; Water = 0.1 ng L^{-1}, Sediment = 0.02 μg kg^{-1} dw, and 12 μg kg^{-1} dw for mussels (Nyberg et al., 2013). In the United States, the Environmental Protection Agency (EPA) limits are set as Aquatic Life Criteria and are assigned as maximum allowable freshwater and salt water exposures for TBT (Freshwater criteria: 0.46 μg L^{-1} acute exposure and 72 ng L^{-1} chronic exposure. Saltwater criteria: 0.42 μg L^{-1} acute exposure and 7.4 ng L^{-1} for chronic exposure).

Within the policy, management of organotin pollution falls under a number of regulations. The Marine and Coastal Access Act 2009 (MCAA) provides the principal statutory means by which the UK complies with EU law, such as the Waste Framework Directive and international obligations such as those under the OSPAR Convention and the London Convention and its 1996 Protocol in relation to disposal at sea. The Marine Management Organisation (MMO) is responsible for most marine licensing in English inshore and offshore waters and for Welsh and Northern Ireland offshore waters and issue marine licenses under the MCAA. Marine Scotland is the licensing authority for Scottish inshore and offshore waters. Natural Resources Wales (NRW) and the Department of Environment Northern Ireland are the licensing authority for Welsh and Irish inshore waters respectively.

This chapter aims to provide an updated review of the chemistry, environmental behaviour, fate, sampling and analysis of organotin compounds. In addition it discusses the overarching policy governing the monitoring and management of these substances in England and Wales.

2. ORGANOTINS AND MARINE BIOTA

Organotin compounds target a large range of organisms, with the biological effects on communities at different locations covered in numerous studies. The most notorious effects relate to organotin induced mortality, shell thickening of oysters, the imposex of gastropods and the decline in the recruitment of marine biota juvenile stages. The first observations of organotin related deformations to non-target (non-fouling) organisms were in 1974, in which French and English oyster planters began noticing abnormal shell thickening and chambering within *Crassostrea gigas*, with observations of shell deformations occurring within the proximity of marinas and boat moorings (Seligman, 1996). Waldock & Thain (1983) and Alzieu et al., (1986) identified the influence of TBT on shell calcification and growth defects, as well as correlations in TBT sea water concentrations and the frequency of oyster shell thickening. Enzyme behavioural response to TBT has been investigated using *Crassostrea gigas*, in which electrophoresis of enzyme allozymes from oyster tissue were measured (Tanguy et al., 1999). The authors found that aspartate-amino-transferase (AAT-2), adenylate kinase (AK) and phosphoglucomutase (PGM) were involved in the main physiological processes affected by TBT and were indicative of pollution within Pacific Oysters. N*ucella Lapillus (*Dog-whelks*)* imposex is one of the most studied organotin toxicological mechanisms; owing to its occurrence at very low ng L^{-1} concentrations (section 5.8 discusses their use in biomonitoring surveys). The sub-cellular mechanism of endocrine disruption is unclear; however organotins are known to induce the inhibition of one aromatase P450, a cytochrome that converts testosterone to estradiol (Graceli et al., 2013). Organotin compounds have demonstrable effects to the reproductive function in mammals (having teratogenic properties) and have been highlighted as obeseogens (chemicals that promote the production of fatty cells or increase the storage of fat) (Junzo et al., 1993; Zuo et al., 2011; Graceli et al., 2013). Microbial responses to organotins vary, with some bacterial strains demonstrating no resistance, whilst others acting as important vectors for adsorption and biological degradation of organotins (Dubey & Roy, 2003). Microbial biofilms consist of complex exopolysaccharides which are frequently anionic in nature and therefore have high propensities for the bioaccumulation of metals and other cations. The bioconcentration factors of biofilms can range over several orders of magnitude (10^{-1} – 10^{-4}) and will be dependent on the accumulating metal/cation, salinity, temperature, and other environmental conditions.

Biofilms act as a sink for nutrients and attract an increasing diversity of planktonic and benthic invertebrates.

The bioconcentration of TBT within these exopolysaccharide have been found to inhibit the natural attachment and metamorphosis of oyster larvae and reef-building corals (Labare et al., 1997; Negri et al., 2001) as well as increase the biomagnifaction risk to grazing invertebrates. Due to the lipophilic nature of tri-substituted organotins, TBT and TPhT are rapidly adsorbed into the lipid structures of marine biota by branchial and dietary uptake. Metabolism rates of organotins within marine invertebrates are generally regarded as being slow; therefore they are easily transported throughout trophic levels (Cima et al., 2003).

A study on TBT uptake, tissue distribution and breakdown in dog-whelks (*Nucella lapillus)*, was undertaken using ^{14}C-labelled tributyltin chloride, in which radiographic measurements of ^{14}C were made on tissue extracts containing total ^{14}C TBT + DBT and TBT (Bryan et al., 1993). Introduction of organotins into animals were undertaken by 1) exposure to seawater, 2) ingestion from a diet of labelled mussels (*Mytilus edulis*) and 3) injection into target regions. Bryan et al., (1993) examined the digestive gland, kidney, mantle, foot, viscera, gonads and blood of the dog-whelks and found that absorption of dissolved TBT primarily occurred via tissues associated with the mantle.

The autoradiographic evidence showed that the ctenidium (gill) and osphradium were probably of most importance in TBT uptake. Of the non-reproductive tissues, the mantle displayed least evidence of TBT degradation, whereas kidney and digestive glands contained the lowest proportions of ^{14}C as TBT than other tissues (60% of proportions found in other tissues). The Dog-whelks fed on labelled mussels reportedly absorbed organotins very efficiently via the digestive gland, which plays a major role in TBT breakdown. Bryan et al., (1993) suggested that the metabolism efficiency was not sufficient in preventing TBT concentration factors of the order of 100,000 (dry tissue/water) within the Dog-whelk.

Organotins affect a whole range of targets in many different ways. Which effect is observed depends on the organism exposed as well as the contaminant concentration. Potentially, TBT acts alternatively or simultaneously as, for example, an enzyme inhibitor, membrane disrupter, radical reagent or a modulator of receptors.

3. Physiochemical Properties, Occurrence, Fate and Degradation of Organotin Compounds

3.1. Physiochemical Properties of Organotins

Organotin compounds are included in the group of organometallics and consist of a central tin ion (typically Sn(IV)) covalently bound to alkyl/aryl organic moieties (Omae, 2003; The World Health Organisation, 2006; Oliveira et al., 2010;). Organotins are anthropogenic in origin and are not considered to be synthesised via natural environmental processes, although bio-methylation of tin has been identified as a fate process within anoxic sediments (Vella *et al*, 2001). The synthesis of organotin compounds can be undertaken by a number of processes (Grignard, Wurtz, alkyl aluminium and direct synthesis) and is followed by a redistribution of the Sn-C bond to the required mono-, di- or tri-alkyl/aryl moieties (Cima et al., 2003). Table 1 presents the chemical structure and the physiochemical properties of mono-, di- and tri-substituted butyl and phenyl tin compounds. The tin organometallic structure is represented as; $R_{(4-n)}SnX_{(4-n)}$where R= the alkyl/aryl organic group, X= the anionic species and n= number of functional groups (Oliveira & Santelli, 2010). Organotin solubility is dependent upon the number of organic moieties attached to the tin atom and has influence over its adsorption rate and partioning behaviour within the environment. Tri-substituted organotins are recognised as the most toxic of these compounds, with TBT and TPhT having the highest toxicity towards many aquatic biota. Toxicity is also attributed to their lipophilic nature; with tri-substituted organotins having the propensity to adsorb on to lipid structures within aquatic organisms. The anionic group (X) is considered as having almost no influence over the toxicity of organotins (unless X is independently toxic); however will have influence over organotin solubility, environmental partitioning and the biological uptake by marine organisms (Looser et al., 1998; Oliveira et al., 2010).

3.2. Occurrence and Fate of Organotin Compounds

By the time the impact to non-target organisms had been documented, tri-substituted organotin compounds had been well established within the coatings market. In 1985, the US Navy calculated annual fuel savings of its 600 ship fleet as $120M, from the application of TBT antifoulants (Champ & Seligman,

1996). The earliest antifouling systems comprised of concentrated tributyltin fluoride (TBTF) and tributyltin oxide (TBTO) manufactured as a blended active ingredient within paint coatings. These were followed by the development of contact leaching paints which released TBT by the gradual ingress of water at the paint surface layer (Anderson et al., 1986; Omae, 2003). In 1974, Milne and Hails patented the first revolutionary "self-polishing" paint (Yebra et al., 2004) allowing the controlled release of TBT at the surface level of the paint, providing the continual renewal of the surface layer. Self-polishing co-polymer paints also provided further economic benefits, with an increase in the interval time between dry docking and re-application of the antifoulant paint (Champ & Seligman, 1996; Yebra et al., 2004). It has been estimated that in Arcachon Bay (France) alone, the use of TBT provoked a loss in revenue of $147M U.S. dollars through reduced oyster production (Readman et al., 1993). Since initial environmental studies, TBT within coatings has undergone increasing restrictions within the UK, with its prohibition delayed through its debate on perceived economic benefits (such as reduced dry-docking costs and carbon emissions). The first regulatory action by the UK government came into force in 1987, with the use of TBT antifoulants banned on recreational boats under 25 m long, as well as limitations on the leaching rates of TBT antifoulants available on the market. On January 1st 2001, the application of organotin based antifoulants was prohibited globally by the International Maritime Organisation (IMO), followed by a complete ban on the presence of organotin antifoulants that act as a biocide in January 1st 2008. Numerous reports document the occurrence of organotins in the environment, generally reflecting the trend that the highest concentrations are found near areas of high shipping activity. Due to busy activity within these regions, potential resuspension and distribution of contaminated sediments from these areas remain a risk factor. Although historical contamination of TBT is designated as the primary source of organotin pollution within the marine environment; other input mechanisms also contribute to their presence. Organotin manufactured within the plastics and glass industry account for the largest consumption of tin globally (The World Health Organisation, 2006) and predominantly use di-substituted and mono-substituted tin compounds as HCl scavengers within the plastic polymer production (Hoch, 2001). The disposal of PVC products and organotin containing commodities to landfill is a contributing factor, with the subsequent leachate emissions of organotins from landfill sites acting as a source of pollution (Björn et al., 2011; Björn et al., 2007).

Table 1. Physiochemical properties of butyltin and phenyltin compounds as the chloride anionic ligand)

Compound	Structural formula	Molecular weight (Da)	Melting Point (˚C)	Boiling Point (˚C)	Solubility* (mg L^{-1})	Log K_{ow}	References
Monobutyltin trichloride ($MBTCl_3$)	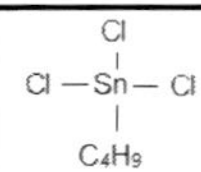	282.8	-63	93 (1.3 kPa)	na	0.18	(Ayanda et al., 2012) (Hoch, 2001)
Dibutyltin dichloride ($DBTCl_2$)	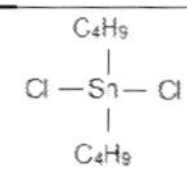	303.8	40	135 (1.3 kPa)	4-50	1.89	(Ayanda et al., 2012) (Hoch, 2001)
Tributyltin chloride (TBTCl)	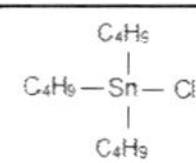	325.5	-16	172 3.3 (kPa)	50	3.54-4.76	(Ayanda et al., 2012) (Hoch, 2001) (Arnold et al., 1997)
Monophenyltin trichloride ($MPhTCl_3$)	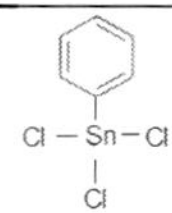	302.2	na	na	na		
Diphenyltin dichloride ($DPhTCl_2$)	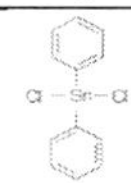	343.8	42	333-337	50	na	(Radke et al., 2008)
Triphenyltin chloride (TPhTCl)	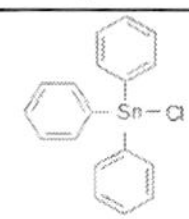	385.5	106	na	1 - 5	4.19	(Ayanda et al., 2012) (Radke, Staniszewska, Wasik, Namieśnik, & Bolałek, 2008) (The World Health Organisation, 1999) (Arnold et al., 1997)

na = not available.
*speciation in sea water.

Table 2. Industrial applications of organotin compounds (adapted from Okoro et al., (2010))

Industrial application	Function	Organotin compound	Reference
Agrochemicals	Fungicide, insecticide	(R = Phenyl) R_3SnX	(Antes et al., 2011)
Antifoulant paint	Biocide application for control of a broad spectrum of organisms	(R = Me, Bu, Oct) R_3SnX	((Hoch, 2001; Omae, 2003)
Glass treatment	Precursor for Sn(IV) oxide films on glass	(R = Me, Bu) Me_2SnX_2, $RSnX_3$	(Buchanan et al., 1997)
Impregnation of goods and application within sanitary products	Fungicide, biocide	(R = Bu, Phenyl) R_3SnX, Bu_2SnX_2	(Yamada et al., 1993)
Industrial water	Biocide in refrigeration and cooling systems	$C_{12}H_{27}Sn$ and $C_8H_{18}Sn$	(Okoro et al., 2010)
Materials protection	Acaracide, fungicide, insecticide,	(R = Bu) R_3SnX	(Omae, 2003)
Poultry farming	Dewormer	(R = Bu) R_2SnX	(Epstein et al., 1991)
PVC stabiliser	Stabilisation/protection from/by heat and light	(R = Bu, Me,) R_2SnX_2 and $RSnX_3$	(Sadiki et al., 1999;Quevauviller et al., 1991)
Wood preservation	Acaracide, fungicide, insecticide	(R = Bu, Phenyl,) R_3SnX	(Fargasova et al., 1997)

Bu = butyl Me =methyl.

Table 3. Concentrations of butyl and phenyl tins in coastal waters (as ng Sn L^{-1})

Location of study	Details of sampling	Analytical method	TBT	DBT	MBT	TPhT	DPhT	MPhT	Reference
Harbours and Bays (15 sites) in South Korea	2009 (August - September) 2010 (September)	GC-FPD (Grignard's reagents)	N.D-23.9	N.D-28.1	N.D-17.9				(Kim et al., 2014)
Port Gdyina, Baltic Sea	2009 (February and June)	PLE-GC-FPD (sodium tetraethylborate)	5.2-79.1	4.2-44.1	1.8-37.6				(Radke et al., 2012)
Ijmuiden, The Netherlands	2010 (August – September)	GC-MS (sodium tetraethylborate)	1.4-1.9	1.3-1.6	0.16-0.29				(Smedes &Beeltje, 2011)
Alicante Harbour, Spain	2005 (November)	GC-ICP-MS (sodium tetraethylborate)	5.8-20	13.3-18.5	2.7-8.1				(Aguilar-Martinez et al., 2008)
Xiamen coast, China	2006 (March)	GC-FPD (sodium tetraethylborate)	1.2-60	0.51-7.3	2.2-67.0				(Wang, Hong, Zhao, & Hong, 2008)
Suruga Bay, Japan	2003 (December) and 2004 (February)	GC-ICP-MS (sodium tetraethylborate)	0.18-13.6	0.09-3.69	<0.04 -3.69	0.01-0.94	0.01-1.2	< 0.01-0.34	(Kurihara, Rajendran, Tao, Yamamoto, & Hashimoto, 2007)
Gíjon region, Spain	-	ID-GC-ICP-MS (sodium tetraethylborate)	0.4-196.6	0.3-33.7	0.8-11.6				(Rodríguez-González, Encinar, Alonso, & Sanz-Medel, 2006)

Concentrations converted to ng Sn L^{-1}. N.D. = not detected/below limit of detection, GC = gas chromatography, FPD = flame photometric detector, ICP-MS = inductively coupled plasma mass spectrometry, PLE = pressurised liquid extraction, ID = isotope dilution.

Table 4. Total sediment concentrations (ng Sn g^{-1}) and sediment pore-water concentrations of butyltins (as ng Sn L^{-1})

Location of study	Details of sampling	Analytical method	TBT		DBT		MBT		Reference
			pore-water	sediment (total)	pore-water	sediment (total)	pore-water	sediment (total)	
Gold Coast, Queensland, Australia	-	GC-MS (sodium tetraethylborate	< 45-1,031	90.7-3,608	< 55-1,692	76.9-2,794	< 70-339	88.1-2,881	(Burton et al., 2004)
Lake Zurich and Lake Constance, Switzerland	Freshwater; concentrations as mean values	GC-MS using perdeuterated standards (sodium tetraethylborate)	7.0	144.3	4.8	56.4	24.4	34.6	(Berg et al., 2001)
Haihe Estuary, Tianjin Harbour, China	-	GC-AAS (Grignard's reagent)	2.8-22.1	11.0-314	6.5-16.4	16.7-212	2.4-22.0	N.D.	(Ma, Dai, & Huang, 2000)
Pandu and Ganda River, India	Freshwater; Pre and post monsoon, 1995	GC-MIP-AES (sodium tetraethylborate)	2.9-32.6		1.7-101.1		2.1-70.1		(Ansari, Singh, & Tobschall, 1998)
Lake Zurich and Lake Lucerne, Switzerland	Freshwater; concentrations as mean values	PTV-LVI GC-MS using perdeuterated standards (sodium tetraethylborate)	9.6		4.5		11.0		(Arnold et al., 1998)

Concentrations converted to ng Sn g^{-1} and ng Sn L^{-1}. N.D. = Not detected/below limit of detection, MS = mass spectrometry, AAS = atomic adsorption spectrometry, MIP = microwave induced plasma, AES = atomic emission spectroscopy PTV-LVI = programme temperature vaporisation-large volume injection.

Table 5. Concentration of butyl and phenyl tins in marine and estuarine sediments (as ng Sn g^{-1} d.w.)

Location of study	Details of sampling	Analytical method)	TBT	DBT	MBT	TPhT	DPhT	MPhT	Reference
Gulf of Gdansk and Vistula and Szczecin Lagoons, Southern Baltic Coast	2008-2009	GC-MS. (Grignard's reagents).	0.2-115.2	< 1.2-38.7	< 1.0-34	N.D.	N.D.	N.D.	(Filipkowska, Kowalewska, & Pavoni, 2014)
Port Camargue	2009 (February) - surface sediments.	GC-ICP-MS (sodium tetraethylborate)	0.2-10,738	0.6-5,066	5.0-710				(Briant et al., 2013)
Elbe Estuary, Germany	2011 (May) - Surface sediments (0-20 cm)	GC-FPD (natriumtetraethyl-borate)	< 0.41-41	< 0.51-10	< 0.68-21				(Wetzel, Winterscheid, & Wahrendorf, 2013)
Elbe Estuary, Germany	2006 (October)	GC-MS	< 0.41-261	1.2 – 231	8.1-339	< 0.34-2.7			(Wetzel, Wahrendorf, & von der Ohe, 2013)
Nine fishing ports, Chinese coast	2007	SPME-GC-FPD (sodium tetraethylborate)	< 0.7-86	< 2.3-42	< 3.6-194			< 5.1-66	(Zhang et al., 2013)
Bitung, Gangga, Island, Manado, and Jakarta Bay in Indonesia	2004	GC-MS (sodium tetraethylborate)	0.16-144	0.46–40	1.0-115	< 0.1-2.4	< 0.1-17	0.12-13	(Harino et al., 2012)
Port Gdyina, Baltic Sea	2009 (February and June)	PLE-GC-FPD (sodium tetraethylborate)	471-2,642	128-1393	91-656				(Radke et al., 2012)
Suez Gulf, Egypt		GC-ECD	114-1,145	56–677			361-920		(Shreadah, Said, Abd El Ghani, & Ahmed, 2011)

Location of study	Details of sampling	Analytical method)	TBT	DBT	MBT	TPhT	DPhT	MPhT	Reference
Port of Gdinya and Port of Gdansk (Baltic Coast)	2008 (November)	GC-MS (Grignard's reagents)	< 0.3-5,780	0.9-2060	< 0.5-684	< 0.5-339	N.D.	< 1-320	(Anna Filipkowska et al., 2011)
San Vincentes Bay, Chile		ID-GC-MS (sodium tetraethylborate).	1024	447	301				(Flores, Bravo, Pinochet, Maxwell, & Mester, 2011)
Hsiao Liouciou Island, Taiwan		GC-FPD (sodium tetraethylborate)	4.0-765	2.4 – 217	13-177	N.D.	N.D.	N.D.	(Liu, Wang, Chung, Leu, & Meng, 2011)
Dunmore East, Ireland		GC-PFPD (sodium tetraethylborate)	2125	790	609				(Giltrap et al., 2011)
São Vicente Estuary, at Pescadores Beach, Brazil	2008 (February)	GC-FPD (Grignard's reagents)	N.D.-66	N.D. – 61	24-161				(dos Santos et al., 2010)
Port of Bilbao, Spain	2007-2008	GC-ICP-MS (sodium tetraethylborate)	0.1-3,143	0.3-2874	< 0.1-125				(Rodríguez et al., 2010)
Zuari Estuary, India	2006 (August)	GC-MS (Grignard's reagents).	6.5-4,353	1.8-1,030	2.0-2238	< 0.2-17	< 0.2-17	< 0.2-2.5	(Jadhav, Bhosle, Massanisso, & Morabito, 2009)
Ports of Menton Garavan, Cap d'Ail, Beaulieu sur Mer, Villefranche sur Mer, Nice and Cannes Port Canto, France	2004 (September)	GC-FPD (sodium tetraethylborate)	37-4,402	34-3,025	65-3,682	2.6-358	< 2-124		(Cassi, Tolosa, & de Mora, 2008)
Venice Lagoon, Italy	2003 (Late spring /summer)	GC-MS-MS (Grignard reagents)	21-39,300	2-12,623					(Berto et al., 2007)

Concentrations converted to ng Sn g^{-1}. DPhT= diphenyltin, MPhT= monophenyltin, N.D = not detected/below limit of detection, GC = gas chromatography, MS = mass spectrometry, ICP-MS = inductively coupled plasma mass spectrometry, SPME = solid-phase microextraction, FPD = flame photometric detector, PFPD = pulse flame photometric detector, ECD = electron capture detector, PLE = pressurised liquid extraction.

Table 6. Summary of passive sampling devices used to monitor organotins

Passive sampling device	Deployment environment and location	Compounds detected	Receiving phase, extraction techniques and analytical procedure	Limits of detection	Reference
Chemcatcher®	Municipal water and Seawater; 14 days - Alicante Harbour, Spain.	MBT, DBT, TBT, TPhT	C_{18} 47 mm octadecyl 3M Empore disk (overlain with a cellulose acetate diffusive layer). Methanol: acetic acid (1:3) into 1 mL hexane. Derivatised using $NaBEt_4$. Analysed by GC-ICP-MS	0.2-7.5 ng L^{-1} (OT cation)	(Aguilar-Martinez et al., 2008)
Semi-permeable membrane devices (SPMD)	Seawater; 15 days - 4 week deployment within Oslo Fjord Harbour, Norway	DBT, TBT	Triolein. 1st: Cyclohexane-DCM. 2nd: Concentrated to 0.5 ml then extracted into 10 mL methanol. 3rd: Extracted into 10 mL hexane with simultaneous derivatisation using $NaBEt_4$. Analysed by GC-AED.	-	(Folsvik et al., 2000)
Semi-permeable membrane devices (SPMD)	Municipal water supply (flow through tank experiment) with other organic pollutants, 28 days.	DBT, TBT, TPhT	LDPE tubing containing triolein. 25 ml of acetate buffer (pH 4.5) added along with 25 ml of hexane. 0.5 mL of $NaBEt_4$ added for derivatisation. Extracts analysed using GC-MS.	-	(Harman et al., 2008)
Silicone Rubber (30 mm x 30 mm x 0.5 mm)	Bottle extractions on deionised and saline water; 6 weeks shaking. Field Trials within Ijmuiden Harbour, The Netherlands	TeBT, TBT, DBT, MBT, TPhT, DPhT, MPhT, TeOT, TOT, DOT, MOT, TeOT, TcHT	Silicone Rubber. Dosing of organotins on to sheets investigated along with 26 extraction techniques. Soxhlet extraction with acetonitrile 1 % acetic acid and a polar solvent (followed by hexane preconcentration) was demonstrated with good recoveries. Analysed by GC-MS.	< 0.1 ng L^{-1} (TBT cation)	(Smedes et al., 2011)

TeBT = tetrabutyltin, TeOcT = tetraoctyltin , TOcT = trioctyltin, DOT = dioctyltin, MOT = octyltin, TeOT = tetraoctyltin, TcHT = tricyclohexyltin, GC = gas chromatography, MS = mass spectrometry, ICP-MS = inductively coupled plasma mass spectrometry.

Table 7. Summary of inland disposal techniques used for disposal and encapsulation of contaminated dredged material

Technique	Method of containment / disposal	Benefits	Potential risks/constraints	UK Status	Cost
Landfill site	CDM pre-treated, dewatered and transported to landfill site.	Used for high level contaminants. Easier access for monitoring. CDM can be mixed with cement matrix to increase stability (see stabilisation / chemical immobilisation).	Potential risk of emissions to air and groundwater. Pre-treatment and dewatering required before disposal. Potential transportation risks associated with waste. Typically hazardous waste landfill required for CDM disposal. Landfill space already at premium.	Yes	Not easily defined. £1-205/t. Port of Bremmen dredged material = approximately 300,000 m^3 per year with cost estimated at £7–10 m per year.
Confined disposal facility (CDF) - upland	Used for high level contaminants. CDM is stored inland and housed within a purpose built facility. Inland CDFs can be lined to prevent leaching of CDM.	Applicable to for high level contaminants. Easily accessible for monitoring. No exposure to hydrodynamic forces.	Potential risk of emissions to air and groundwater. Potential transportation risks associated with waste. Limitations over sediment type - not all CDM can be placed in one repository. Potential viable site limitations	Limited by license	£8-59/m^3 + dewatering + transport+ monitoring costs
CDF-Near-shore	Impermeable walls (dykes) at shoreline used for the deposition of CDM.	Disposal facility can be placed in close proximity to dredging operations. In some circumstances can be utilised as recreational /wildlife reserves.	Potential risk of emission to air and groundwater. Limitations over sediment type - not all CDM can be placed in one repository. Possible negative impacts to coastline and navigation routes.	No	-

Table 7. (Continued)

Technique	Method of containment / disposal	Benefits	Potential risks/constraints	UK Status	Cost
CDF - Silt Lagoons	On-site inland lagoons excavated for the deposition of CDM in the vicinity of dredged material removal. Can be lined with clay or plastic.	Potential development of nature reserves and/or recreational facilities after backfilling complete.	Potential risk of emission to air and groundwater. Limitations over sediment type - not all CDM can be placed in one repository.	Limited by license	-
Stabilisation /chemical immobilisation	CDM is chemically mixed with cement, lime, calcium, aluminates or fly ash to immobilise pollutants.	Makes CDM more inert. Production of secondary construction materials (bricks, gravel, artificial basalt). Reduction of space taken up at disposal sites.	High energy through thermal stabilisation. Implementation and management of erosion control to reduce dust and nutrient emissions.	Yes- if deemed for beneficial use	Dependent on production material; Costs range £11-52/m^3.

Table 8. Summary of sea disposal techniques used for disposal and encapsulation of contaminated dredged material

Technique	Method of containment / disposal	Benefits	Potential risks / constraints	Undertaken in UK	Approximate Cost
Contained Aquatic Disposal (CAD)	Normally undertaken in low energy environments. Dredged material is disposed of within a designated excavated basin on the sea bed. Basin capped using inert marine or terrestrial muds once capacity is reached. Geosynthetic layers can be introduced to reduce contaminant emission.	Excavated near to dredging area. Excavated area can offset cost through aggregate extraction. Less handling/ transportation of contaminant	Requires monitoring and potential reestablishment of the capping layer. Disturbance from other sea users (e.g. trawlers) can present a disturbance risk. Bioturbation from burrowing organisms highlighted as another potential route for contaminant remobilisation. Implication of inclusion/exclusion zones on sea bed. Until capped, long term exposure risk to surrounding environment. Loss of CDM during placement in deep waters (> 20 m).	Limited by license	£4-26/m^3
Island CDF facility	Constructed off shore repository for dredged material in which dredged wastes are placed within. CDF walls are constructed higher than surrounding water level and are capped once at capacity.	Constructed near to dredging area. Easier to monitor than sea bed CAD sites. Isolated at base and top.	Potential risk of emissions to air and surrounding water. Slower biological reduction of contaminant. Potential of spill during CDM deposition in CDF facility. Subject to hydrodynamic influences.	No	-

Table 8. (Continued)

Technique	Method of containment / disposal	Benefits	Potential risks / constraints	Undertaken in UK	Approximate Cost
Open sea disposal with capping	CDM deposited within a low energy environment and capped using inert silt/sand substrate. Potentially viable for the retention of CDM at lower contamination levels. Reduced transport costs and simple disposal technique.	Simple disposal technique located close to port/harbour. Reduction in energy use, transport and use of space. Capping layer can be replenished by ongoing maintenance dredging operations.	CDM is exposed until the time of capping, having implications of persistent pollutant release to surrounding environment. Capping layer requires constant monitoring and potential re-establishment with fresh material. Implications of inclusion/exclusion zones on sea bed.	No - Trial in progress at North Tyne	£33/m^3 + monitoring costs
Geotextile bags	CDM placed within flexible tube units and disposed of out to sea. CDM separates contaminants from surrounding environment	Placed in close proximity to dredging operations. Reduction in transport costs. Used to construct underwater structures. Can be used as infill for construction projects.	Potential damage to bags causing emissions to surrounding environment. Implications of inclusion / exclusion zones to sea bed. Additional handling of CDM	Limited by license	£80/m^3

Other emissions to the environment include leaching from newly installed PVC water pipes (Quevauviller et al., 1991; Sadiki et al., 1999; Wu et al., 1989) and emissions from sludge waste in sewage treatment facilities. Tri-substituted organotin compounds are still used as agricultural pesticides within certain regions of the world, notably with use of triphenyltin (TPhT) and to a lesser extent, tripropyltin (Cima et al., 2003). Phenyltin (fentin) is applied in the form of TPhT acetate and TPhT chloride and is a contact pesticide for protection of crops against fungal diseases. TPhT was introduced into the international market in 1960 but has now been excluded from Annex 1 of the EC Directive 1107/2009 (IUPAC, 2011). TPhT was also historically introduced as a co-toxicant to TBT antifoulant paints. A summary of industrial applications of organotin compounds are provided in Table 2. Figure 1 presents a simplified overview of the environmental fate process of current and historical inputs, outlining the fate processes for organotin compounds.

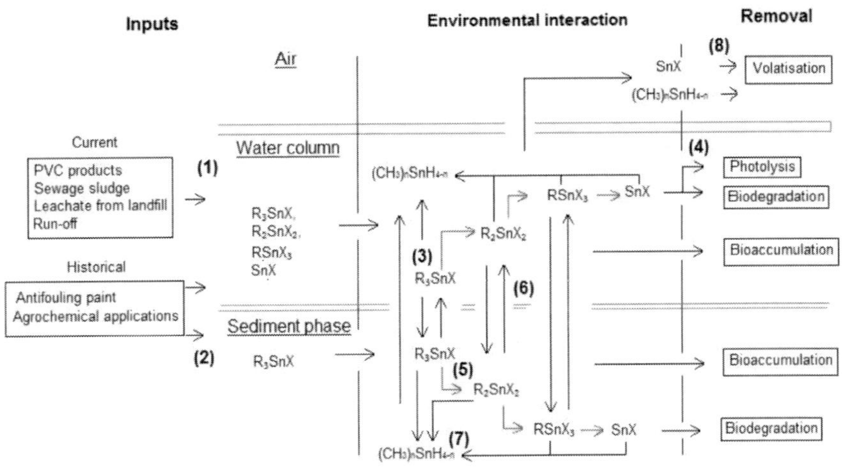

Figure 1. Environmental fate pathways of organotin compounds in the marine environment.

Current or "new" inputs of organotins to the marine environment typically source from the contamination of municipal water occurring from the leaching of organotins in sewage sludge, landfill sites, industrial dispersal and PVC pipes/plastics (1). Inputs from historical contamination are sourced from TBT and TPhT antifoulant paints, which can be present as sediment bound concentrations or as paint particulates still present within sediment (2). TBT half-lives within sea water are reported in the order of days to weeks, with concentrations in the water column typically much lower than in sediments, owing to the rapid adsorption to the particulate phase (particularly for tri-substituted species) (3). The highest concentrations of organotin compounds within the water column can be found at the surface microlayer, where photolysis exhibits a degradative role (4). Half-lives of TBT within sediments have been reported as several years. Within this compartment removal by bio-degradative processes are of most importance (5) (particularly as a large majority of sediments occur below the photic zone). Desorption readily occurs and is thought to occur from mineralisation of sediments and the hydrolysis of cationic binding ligands on the sediment surface (6). Biomethylation of organotins and their respective fluxes have not been widely reported but are thought to contribute as a fate process, with the addition of methyl groups to the tin atom occurring most within anoxic sediments (7). Methylated organotins have a lower affinity for the sediment phase in comparison to non-methylated species and will therefore have a higher tendency to desorb (Vella et al., 2001). Losses of methylated TBT from dredged sediments to the air have also been reported (Vella & Vassallo, 2002). Volatisation of organotin compounds can be accounted as a removal process from the environment (8); however are not considered to be a major mechanism, with the half-life for volatisation of TBT from water reported as 11 months by the World Health Organisation (1990).

3.3. Partitioning of TBT and TPhT in the Aquatic Environment

Approximately 95% of TBT in the water column is bound to suspended particles, with the remainder largely associated with dissolved organic matter and organic/inorganic ligands (Gadd, 2000). Benthic sediments act as the major sink for TBT (Berg et al., 2001; Hoch et al., 2002; Burton et al., 2004; Antizar-Ladislao, 2008), where the adsorption of TBT and TPhT to the sediment/particulate phase is a dualistic function of hydrophobic partitioning and electrostatic interaction; governed by the pH and the salinity of the aquatic environment (Arnold et al., 1997; Burton et al., 2004; Hoch et al., 2002). The aquatic chemistry and the subsequent organotin-ligand speciation also have an influence on the bioavailability and bioaccumulative properties of organotin compounds, with the potential for accumulation into lipid tissue assessed using the *n*-octanol water coefficient (K_{ow}) (Arnold et al., 1997; Bangkedphol et al., 2009).

The environmental factors influencing the behaviour of organotins into sediment/particulate phases include:

1) The sediment/particulate composition; organic carbon content, particle grain size and the availability of deprotonated anionic binding sites (or the net negative charge) of the solid-phase.
2) The number of organic substituted groups (R) covalently bound to the tin atom; e.g. mono-, di- and tri–moieties (e.g. butyltin species = MBT, DBT, TBT).
3) Organotin-ligand or cationic complexation; as a function of aquatic pH, salinity and the presence/nature of anionic ligand species present within the system. For example, TBT and TPhT ionic speciation (TBT^+ and $TPhT^+$) occur below their acidic constants of 6.25 and 5.20 respectively (Arnold et al., 1997) and will therefore have increasing electrostatic interactions with the tin atom and particulate/sediment surface ligands (Berg et al., 2001). In marine environments, TBT and TPhT are speciated predominantly as neutrally charged TBTOH complexes with partitioning behaviour being more typical of a hydrophobic organic contaminant (Brandli et al., 2009).

Investigations into the adsorption behaviour of TBT on to the solid-phase look to quantify interactions using the solid-water partition coefficient (K_d), in which a linear relationship between the mass of chemical in the sorbed and aqueous phase are at equilibrium (normally it is assumed to be a Freundlich adsorption isotherm). Bangkedphol et al., (2009) measured the K_d of TBT on three sediments of differing properties with varying aquatic conditions and compared these data against computational models. They found that adsorption was highest at pH 6.5, with adsorption of TBT greatest within fresh water. Adsorption decreased at low and intermediate salinities (25–50 psu) and was then found to increase again at higher salinities (> 50 psu), with sediment organic carbon concentration increasing the adsorption of TBT (Bangkedphol et al., 2009; Unger et al., 1996).

3.4. Desorption and Pore-Water Resupply of Organotin Compounds

A limited number of studies have investigated partitioning of organotins in sediment pore-waters. As the extractable quantities pore water can be low, studies assessing adsorption and desorption kinetics of organotins within this compartment have used specialised gas chromatography (GC) injection techniques, perdeuterated internal standards (Arnold et al., 1998) and/or carried out assessments on elevated concentrations obtained from highly contaminated sediments (Burton et al., 2005).

Berg et al., (2001) introduced sorption/desorption kinetic model (as the organic carbon-normalised-sediment water distribution ratio= D_{oc}) for organotin compounds in two fresh water harbour sediments in Switzerland. Experiments were undertaken in three stages and examined contaminated sediment and the behaviour of perdeuterated labelled organotin standards (e.g. TBT-d_{27}). Berg et al., (2001) suggested that as a proportion of organotin sorption to sediments within fresh water (at pH 7.3) is predominantly undertaken by electrostatic interaction to sediment surface ligands, in which the polar properties of sediment bound oxygen ligands are more easily accessible than the hydrophobic domains, allowing for the reversible desorption mechanism of organotin compounds to occur. Burton et al., (2005) investigated the adsorption/desorption kinetics of heavily contaminated sediment cores from an estuarine marina in South East Queensland, Australia. In this study, the vertical partitioning behaviour of butyltins between total sediment concentrations and pore-waters was assessed. Burton et al., (2005) reported that within the first 0-6 cm of the sediment profile, a higher variability of D_{oc} was observed; with an increasing concentration gradient of butyltins with depth. They suggested that the higher D_{oc} variances in this region of the core were not at equilibrium, indicating the potential upward diffusion of butyltins from > 6 cm depth to the sediment-water interface. Using the butyltin degradation index (BDI); (see section 2.3.1.), they also found that values < 1 were observed within the first 0–2 cm of the core, indicating a potential fresh input of TBT from the previous 1.3–8.7 years. Concentrations of degradation products (DBT and MBT = BDI of 2.53) were higher in sediments > 10 cm depth, suggesting no recent inputs of TBT and a natural attenuation of historical TBT by overlying sediments. Data from these studies show that sediments can potentially encapsulate and act as a repository for TBT, where they will be retained until fully degraded to inorganic tin. Due to the susceptibility of organotin compounds to desorb from the solid-phase, sediments act as a source of pollution, in particular from their disturbance and resuspension. Boating and dredging activities have therefore been identified as a remobilisation mechanism of organotins to the water column.

3.5. Degradation of Organotin Compounds

Half-lives of TBT within the following aquatic compartments have been reported as; freshwater = 6–26 days, estuarine water = 1–2 weeks, sea water = 6–127 days, estuarine sediment = 3.8 years and marine sediment = 1.85–8.7 years (Cima et al., 2003). Degradation of organotin compounds occur by the step-wise loss of the organic group, successively degrading to inorganic tin. The organotin degradation route can be expressed as;

$$R_3SnX \rightarrow R_2SnX_2 \rightarrow RSnX_3 \rightarrow SnX_4$$

(Blunden & Chapman, 1982)

e.g. for the degradation of TBT:

TBT $\rightarrow$ DBT $\rightarrow$ MBT $\rightarrow$ Sn(IV)

Abiotic degradation is governed by water pH, salinity, temperature, exposure to light and the solubility of organotin compounds (Dubey & Roy, 2003). The rate of degradation is also governed by algal or bacterial communities present within the system and their respective concentrations within the environment (Dubey & Roy, 2003). Thermal degradation of organotins is often discounted as an environmental mechanism, due to the Sn-C bond having stability at temperatures > 200°C (Seligman et al., 1996). Gamma irradiation has also been identified as having negligible effect on degradation rate, as radiation energies required for cleaving the Sn-C bond are reduced from the attenuation of the Earth's atmosphere (Seligman et al., 1996). Degradation by hydrolysis is an abiotic mechanism, however, is only achieved under extreme pH conditions, rarely encountered within the natural environment (World Health Organisation, 1990). Sunlight within the ultra-violet spectrum has energy of 600 kJ mol^{-1}, and is considered to have a slight influence over the cleaving of Sn-C bonds (with disassociation energies ~ 210 kJ mol^{-1}) (Blunden et al., 1982). Biotic degradation due to microbial activity is more rapid in the water column due to higher light intensities and greater oxygenated conditions in comparison to sediments (Radke et al., 2008). Biotic degradation processes are considered the most important mechanism within the environment, with butyltin degradation occurring at low concentrations from interactions with algae and bacteria (Gadd, 2000). In general, the toxicity of TBT and TPhT to marine organisms have been reported to be reduced with the successive removal of the organic moiety (Paton et al., 2006); although the toxicology of organotins is complicated due to variances in environmental concentration, exposure time, compounds present *in-situ* and the overall sensitivity of biota to organotin species (Graceli et al., 2013).

3.6. Butyltin and Phenyltin Degradation Index

The BDI and the phenyltin degradation index (PhDI) are used to predict recent inputs of TBT or TPhT contamination into sediments. By calculating concentration ratios of the degradation products against the concentrations of the primary TBT or TPhT compound, the BDI or PhDI value is obtained. Values < 1 are indicative of recent inputs to sediment; whereas values which give a ratio > 1 are representative of organotin degradation, with no "recent" inputs of TBT or TPhT.

1) $\mathrm{BDI} = \frac{[\mathrm{MBT}]+[\mathrm{DBT}]}{[\mathrm{TBT}]}$

1) $\mathrm{PhDI} = \frac{[\mathrm{MPhT}]+[\mathrm{DPhT}]}{[\mathrm{TPhT}]}$

(Díez et al., 2002)

4. Concentrations of Organotin Compounds in Marine Compartments

Since restrictions were imposed in 2003 and its total prohibition in 2008, overall total concentrations of TBT within marine sediments have reduced, with concentrations of organotins found within environmental compartments largely dictated by local and historical situation. Due to the partitioning behaviour and low solubility of TBT and TPhT, concentrations within the water column (Table 3) are much lower than total sediment concentrations (Huang et al., 1998). Despite sediments acting as the major sink for organotin compounds, there is a paucity of studies providing pore-water concentrations (Table 4). Total sediment concentrations (Table 5) can vary from levels below detection to highly contaminated ($\mu g\ g^{-1}$). Concentrations are reportedly higher during warmer seasons and have been attributed to an increase in boating activity and biological processes (with increased degradation of butyltins and desorption from remineralisation of organic matter) (Radke et al., 2012).

When comparing concentrations (Table 5) with OSPAR EAC sediment guideline level (10 ng TBT kg^{-1} dw), the majority of TBT concentrations fall below the guideline values (after conversion back to TBT cation concentrations), although the large variation of TBT concentrations for each site are indicative that potential localised hotspots amoungst sampling locations for each study may be present.

Water concentrations for TBT in Table 3 are in excess of the EU WFD limit of 0.2 ng L (after converting back to cation concentrations). Due to the limited number of recent studies, it is not possible to describe the overall status of organotins in coastal waters. The elevated concentrations in water may suggest that despite declining sediment concentrations, redistribution of TBT to the water column from sediments along with other inputs may still be occurring and that resuspension fluxes may have a greater resupply influence than previously thought.

5. Sampling and Analysis of Organotins in the Aquatic Environment

5.1. Sampling

Sampling of organotins in water and sediment is generally undertaken using amber glass bottles, although polycarbonate, PTFE and aluminium can also be used (OSPAR Commision, 2013). PVC materials should be avoided as they can potentially leach di-substituted butyltins (Quevauviller et al., 1991). Once collected, samples must be kept cool during transportation with short-term storage of sediments optimally at 4°C. If long-term storage is required, freezing at -20°C has been found to be suitable, with samples being stored for up to one year (Gomez-Ariza et al., 1999; OSPAR Commision, 2013). For water samples, non-acidified samples stored in the dark at 4°C in Pyrex glass bottles have been found to be suitable for up to 5 months (Leroy et al., 1998). The long-term stability of butyltins has been reported to be effected with samples containing high concentrations of suspended matter, specifically due to interactions with particulates and microbial communities contained within the sample (Leroy et al., 1998).

5.2. Extraction of Organotin Compounds from Sediments

As sediment heterogeneity has a large influence over the partitioning of organotins, physiochemical parameters of the sample (e.g. total organic carbon % (TOC), grain size and carbonate content) are usually determined in conjunction with organotins. Internal standards (e.g. tripropyltin or tetrabutyltin) are crucial when undertaking the assessment of overall organotin recoveries from extraction (Smedes, et al., 2000) and are used to assess evaporative losses or inefficiencies with the derivatisation procedure. Validation of an analytical procedure requires optimal recoveries on certified reference materials (CRMs). The two most commonly used CRMs for butyltins and phenyltins are PACS-2 (from the National Research Council Canada; for marine sediment which contains MBT, DBT, TBT) and BCR-646 (from the European Commission Joint Research Centre; for freshwater sediment containing MBT, DBT, TBT, MPhT, DPhT, TPhT). Homogenisation of sediments prior to extraction should be undertaken, particularly as paint particulates can be present within samples. Sample sizes between 1–5 g are generally used for extraction with the choice of sediment extraction method dependent upon the range of organotin species of interest. Tri-substituted organotins require extractions using a medium-to-low polar solvent, owing to the lipophilic nature of the trialkyl/aryl tins. For less organically substituted compounds, such as MBT, in which the influence of mineral/electrostatic binding is greater, treatment with an acidic solvent mixture can be used (e.g. acetic acid:methanol) (OSPAR Commision, 2013). The use of complexing agents (such as tropolone or carbamates) can be used to increase the extraction yields from the aqueous to organic phase, notably for the lesser substituted compounds (e.g. MBT) (Antizar-Ladislao, 2008). Alkaline saponification has also been used for extracting organotins from sediments (Bolam et al., 2014). Mechanical shaking, microwave assisted extraction, supercritical fluid extraction, accelerated solvent extraction can all be used as the extraction techniques. The use of deuterated organotin compounds as internal standards are used to improve analytical limits of detection, as well as allow for critical evaluation of changes and degradation processes occurring during the analytical procedure (Brunori et al., 2006). The use of ^{119}Sn enriched standards have been incorporated with mass spectral techniques for the analysis of butyltins in aqueous and sediment samples (Centineo et al., 2004). For this technique, known amounts of enriched isotopic standard are spiked to samples and the percentage response ratio against the natural tin ion is calculated. This approach can then be used to quantify butyltins in biota, sediment and water samples (Centineo et al., 2007). To the best of knowledge, there are currently no isotopic enriched standards for phenyltin compounds.

5.3. Derivatisation

For gas chromatography (GC) applications, it is necessary to convert organotin species into volatile, thermally stable derivatives. Alternative approaches include hydrogen bromide column doping techniques (Mizuishi et al., 1998), however, are not often favoured due to the destruction of the active sites within the GC capillary column.

Organotin compounds are typically converted into;

1) Alkyl derivatives, using Grignard reagents (such as hexylmagnesium bromide) or ethylation using sodium tetraethylborate ($NaBEt_4$).
2) Hydride species (e.g. using sodium borohydride ($NaBH_4$), converting organotins to hydride derivatives.

5.3.1. Grignard Reagents

A variety of Grignard reagents (RMgBr) are available, so the volatility of the derivatised organotin compounds can be manipulated by the degree of the conversion alkyl group. For practical reasons, pentyl- or hexyl-magnesium bromides are typically used (OSPAR Commision, 2013). Optimum alkane length is important as loss of analyte can occur in the extraction and work-up procedure with volatile short alkane chain derivatives. Conversely, for longer chained carbon groups, long GC retention times can occur. Grignard reagents are hazardous chemicals and react violently with water, acids, ketones and alcohols (Morabito et al., 2000), therefore, very dry conditions and expert laboratory handling techniques are required for their use. Derivatisation using Grignard reagents can only be undertaken in non-polar solvents. Morabito et al., (2000) reported that the Grignard method provided the greatest yields and reproducibility in all common environmental matrices, when compared to sodium tetraethylborate and sodium borohydride derivatisation. Excess Grignard reagent can be destroyed by adding acid (hydrochloric or sulphuric) or saline (ammonium chloride) solution (Morabito et al., 2000). The alkylation reaction of organotin by a Grignard reagent can be expressed as:

$$R_xSnX_{(4-x)} + \acute{R}_xMgX \rightarrow R_xSn\acute{R} + H_2$$

(Al-Rashdi, 2011)

(R = organic substituents and X = anionic group).

5.3.2. Alkylation with Sodium Tetraethylborate

Alkylation with sodium tetraethylborate (often referred to as STEB) allows for organotin compounds to be converted into their ethyl derivatives by nucleophilic substitution (Morabito et al., 2000). Derivatisation is pH sensitive (with optimum ranges reported at 4–6). Lower pH ranges causes the decomposition of the STEB compound due to reactions with the H^+ ions, whilst at higher pH ranges, organotins take on hydroxide speciation (Smedes et al., 2000). The introduction of an acetate buffer solution is usually added to samples to regulate pH during the derivatisation process (Morabito et al., 2000; Smedes et al., 2000). Derivatisation with STEB can be used either simultaneously during or after most of the commonly used aqueous phase extraction techniques. STEB reagent is often made up to concentration of 1–5 %, within deionised water or methanol (Smedes et al., 2000). STEB does, however, have a relatively short shelf life (~ 3-4 days in deionised water at 4°C kept in the dark). Reagent life time can be extended by freezing at -20°C, although most analytical methods suggest that a fresh batch is made for each set of extractions (Smedes et al., 2000). The ethylation reaction of organotin by sodium tetraethylborate can be represented as:

$$R_xSnX_{(4-x)} + NaB(C_2H_5)_4 \rightarrow (C_2H_5)_xSnR_{(4-x)}$$

(Al-Rashdi, 2011)

5.3.3. Hydride Generation Using Sodium Borohydride

Using an aqueous solution of sodium borohydride, organotin compounds can be converted into more volatile hydride derivatives. Methods can include off-line procedures in which simultaneous extraction and derivatisation into a non-polar solvent is used (Jiang et al., 1996). Sodium borohydride has typically been used with on-line procedures such as purge and trap and cryogenic U-tube trapping atomic adsorption spectrometry (Martin et al., 1994) (see section 5.4). Derivatisation using sodium borohydride can suffer from interferences within complex matrices such as sediments and biota; mainly from the interaction with metals and the subsequent production of metal borides (inhibiting the formation Sn-H bonds on organotin compounds) (Morabito et al., 2000). The hydride derivatisation reaction of organotins using sodium borohydride is represented as:

$$R_xSnX_{(4-x)} + NaBH_4 + H^+ \rightarrow R_xSnH_{(4-x)} + H_2$$

5.4. Gas Chromatographic Separation Techniques

GC is the most widely used technique for the separation of organotins due to its high resolving power and the wide range of detectors available. Modern GC methods are able to separate organotins of differing organic moieties (butyl, phenyl, propyl) throughout one sample run (Oliveira et al., 2010). The earliest techniques include U-tube cryotrapping and heating. With this technique, organotin compounds are leached from an acid solution, derivatised and purged from the reaction vessel using an inert gas, followed by their pre-concentration on to a chromatographic sorbent (Martin et al., 1994). Analysis using this method has typically been undertaken on leachates (Smedes et al., 2000) and coupled with atomic adsorption instruments (such as quartz furnace or electrothermal atomisation). Modern approaches for the separation of derivitised organotin compounds generally use non-polar GC capillary columns (e.g. 5% phenyl-methylpolysiloxane stationary phase) (Smedes et al., 2000). Typically, splitless injection (250–280°C) with the oven temperature held below the boiling point of the extractant solvent is used. Other injection techniques (e.g. programmed temperature vaporization large volume injection) can be used to improve limits of detection. In this application larger sample volumes are injected into a cooled inlet temperature (held just below the boiling point of the extractant solvent) and vented off before the rapid heating of the inlet. Tin selective detectors that can be used with GC analysis include flame photometric detection (fluorescence of tin atoms using cut-off interference filter), atomic emission detector, and mass spectrometry (where the selection of 2-4 representative ions are used for each organotin compound; (commonly referred to as selective ion monitoring or SIM)) (Smedes et al., 2000). The most sensitive detection techniques are able to achieve instrument detection limits (IDL) in the low pictogram (pg) to femtogram (fg) range. These include pulse flame photometric detector (Leermakers et al., 2005), tandem mass spectrometry (Tsunoi et al., 2002) and inductively coupled plasma mass spectrometry (Wahlen et al., 2002).

5.5. High Performance Liquid Chromatographic Separation Techniques

For high performance liquid chromatography (HPLC) derivatisation is not required, allowing for a faster analytical procedure and eliminating a potential source of cross-contamination and potential discrepancy in derivatisation yields. Limitations include restrictions in the sensitivity on the types of detector that can be coupled to HPLC, as well a limited range of organic speciation possible with one running sequence. The most commonly used detectors are mass spectrometry, tandem mass spectrometry, inductively coupled mass spectrometry and fluorimetry (González-Toledo et al., 2003). A variety of mobile and stationary phase combinations have been used for the analysis of organotins. These include ion-exchange, reversed phase, normal phase, ion-pair, size exclusion, micelle and vesicle-mediated and supercritical fluid systems (Oliveira et al., 2010). Stationary phases can broadly be categorised by either ion-exchange or reversed phase chromatography (Harrington et al., 1996; Oliveira et al., 2010). Ion-exchange stationary phases are typically made from styrene divinylbenzene resin or silica (Harrington et al., 1996; Oliveira et al., 2010), in which cationic organotin species compete with the mobile phase counter ions for ionic sites (Harrington et al., 1996; Oliveira et al., 2010). Styrene divinylbenzene resin suffers the effects of swelling from the mobile aqueous phase and causes its compression. Cross-linking can overcome this; however, causes a decrease in the mass transfer process (Harrington et al., 1996; Oliveira et al., 2010). Silica groups are more stable allowing for use of faster elution and high column pressure conditions. Silica columns are pH sensitive and are only stable at pH ranges 2–8 therefore buffer solutions are used to reduce peak tailing (Harrington et al., 1996; Oliveira et al., 2010). With ion-exchange chromatography, the strong retention of the mono-substituted organotins to the stationary phase can be problematic, and the use of complexing agents and pH gradient elution are often needed (González-Toledo et al., 2003). For silica-based columns, most mobile phases use methanol (50–90% v/v) with an added salt (e.g. ammonium acetate or citrate at 0.005–0.2 mol L^{-1}) (Oliveira & Santelli, 2010; Harrington et al., 1996). With reversed phase chromatography, a polar mobile phase is used in conjunction with a non-polar stationary phase (such as an alkane group e.g. C_{18} = octadecyl, attached to an end capped silica support material) (Smedes et al., 2000; Oliveira et al., 2010). The eluent is typically water mixed with an organic modifier such as methanol, where the elution strength is increased over time (Oliveira et al., 2010).

5.6. Electrochemical Detection Techniques

Use of electrochemical techniques for the detection of organotins has received notable interest. The development of a low-cost, portable, easy to use detection device can provide benefits to environmental regulators and port authorities for their monitoring and enforcement requirements. For TBT, electrochemical techniques, (in particular voltammetric methods) have been widely used for the detection of free metals, and in some cases organometallic complexes in environmental samples. Difficulties with this detection approach concern the relatively low concentrations of organotins in the marine environment, together with the existence of several organotin compounds and the interferences from other metals (Ferri et al., 1994). Adsorptive cathiodic stripping voltammetry techniques for inorganic Sn utilise hanging mercury drop electrode along with complexing ligands such as tropolone, cathecol, oxine and chloranilic acid (Heppeler et al., 1996; Gao et al., 1996; Adeloju, 1991; van den Berg et al., 1989). Differential pulse polarography methods have been used to measure butyltin species at ng L^{-1} or ng g^{-1} concentrations. This approach, however, requires extensive sample pre-treatment and pre-concentration (Ferri et al., 1994; Weber, 1986).

5.7. Passive Sampling Techniques for Monitoring Organotin Compounds in Water

For the analysis of organotins, large volumes of water are often required for use in conventional analytical methods in order to achieve the required low limits of detection. Use of passive sampling devices is an alternate approach where the deployment of *in-situ* devices and can provide lower limits of detection and can effectively sequester larger volumes of water over time. After deployment, conventional extraction methods are used to determine pollutant concentrations within the sampling device. Depending on deployment time, these passive samplers can provide time-weighted average (TWA) concentrations of freely dissolved pollutants in a water body, however, due to suspended particulates and other interacting material, will not account for whole water concentrations (Smedes et al., 2011). The use of passive sampling techniques for the measurement of organotin compounds have received increased attention over the last decade. This is primarily due to organotins being present in water at inherently low environmental concentrations (~ ng L^{-1}) and the low level limits set be the EU Water Framework Directive (0.2 ng L^{-1}). Butyl and phenyltin compounds have log K_{ow} values of 4.10 for TBTOH and 5.20 for TPhTOH (Arnold et al., 1997) and therefore have potential for accumulation within apolar passive sampling device (Smedes et al., 2011). Table 6 presents organotin passive sampler studies. The organotin Chemcatcher® houses a C_{18} 3M Empore® solid-phase extraction disk within a PTFE sampler housing. The C_{18} disk is overlain with a cellulose acetate diffusion-limiting membrane. The sampling rate of the Chemcatcher® is determined from calibration tank experiments. Aguilar-Martinez et al., (2008) determined the organotin Chemcatcher® sampling rates using temperature and flow rate as the variables; MBT = 6–18 mL day¯1, DBT = 41-204 mL day^{-1}, TBT = 29-202 mL day^{-1}, and 26-173 mL day^{-1} for TPhT. Field deployments of the Chemcatcher® sampler were undertaken at Alicante harbour for a 14-day sampling period. TWA concentrations for field samples were; MBT = 12 ng L^{-1}, DBT = 27 ng L^{-1} and TBT = 28 ng L^{-1}, along with a BDI of 1.4, indicating no recent input of TBT (Aguilar-Martinez et al., 2008). Sampling rate and limits of detection for organotin passive sampling have been further improved using silicone rubber sheets, notably due to the larger exposure surface area of the sampler (300 cm^2). For butyltins, the more hydrophobic compounds with the highest Log_{kow} demonstrated the fastest sampling rates with MBT = ~1.0 L day^{-1}, DBT = ~1.8 L day^{-1} and TBT = ~2 L day^{-1} with limits of detection for TBT in the order of 0.1ng L^{-1} (Smedes et al., 2011). Smedes *et al* (2011) found that diffusion coefficients of organotins were relatively slow, with di-substituted organotins having the lowest diffusion coefficients. Diffusion coefficients for ionic species were much lower than neutrally associated organotins (Smedes et al., 2011). Studies using semipermeable membrane devices (SPMD) have incorporated triolein within a low-density polyethylene (LDPE) membrane (Harman et al., 2008; Folsvik et al., 2000). Harman et al., (2008) report sampling rates as TBT = 2.67 L day^{-1}, DBT L day^{-1} = 1.63 and 0.84 for TPhT. MBT was not detected within the SPMD sampler. Although passive samplers can provide lower levels of detection through extended sampling times, biofouling of samplers and the influence of the diffusive boundary layer at the water-sampler interface (as a function of flow rate variations) can cause inaccuracies with TWA concentrations. These should therefore be considered as part of the passive sampler validation and monitoring procedure. Depending on deployment time, passive sampling devices can also allow for detection of intermittent pollutant release events, which can be missed with conventional spot sampling.

5.8. Biomonitoring Strategies for Organotin Compounds

Biomonitoring techniques can demonstrate a direct link between a contaminant and the response of an ecosystem to its presence. Ideally the organism should accumulate contaminants at a sufficiently slow rate to allow the tissue concentrations to reflect mean environmental concentrations occurring over a short period of time and not respond rapidly to short-term fluctuations (Waldock et al., 1986). Primary investigations which first highlight the impact of organotins to the marine environment are presented in biomonitoring studies of shell thickening within French and UK Pacific Oyster (*Crassostrea gigas*) fisheries and the TBT induced reproductive failures brought on by imposex of Dog-whelks (*Nucella lapillus*) (Waldock et al., 1983; Alzieu et al., 1986; Gibbs et al., 1986). Mollusc sensitivity to TBT and TPhT is particularly high, with the imposition of male characteristics in female snails occurring at low concentrations (ng L^{-1}) (Graceli et al., 2013). The severity of endocrine disruption occurs in a dose-dependent manner making imposex of molluscs a good long-term bioindicator for TBT and TPhT (Graceli et al., 2013). Imposex indices have been used for a wide range of aquatic biota in organotin biomonitoring studies and include; the Relative Penis Length Index (RPLI), Relative Penis Size Index (RPSI) and the vas deferens sequence index (VDSI) (Galante-Oliveira et al., 2010; Horiguchi et al., 2012). Biomonitoring techniques cannot determine the current concentrations of a pollutant; however data from integrated studies of TBT sediment contamination and imposex of gastropods have determined that TBT concentrations can be estimated using these indices values. Variations in specimen age and the season of sampling is an important consideration when undertaking biomonitoring surveys. OSPAR and the International Council for the Exploration of the Sea (ICES) have developed a number of Ecological Quality Objectives (EcoQOs) as part of a North Sea pilot project, in which 10 biological indicators are used to assess management activities and environmental health of the North Sea region. Imposex in Dog-whelks (*Nucella lapillus*) is one EcoQO and is included as a mandatory component of the OSPAR Co-ordinated Environmental Monitoring Programme (CEMP) (CEMP Assessment Report., 2009) in which the spatial distribution and imposex data of gastropod species populations in the North Sea are collated. CEMP guidelines use the VDSI to assess imposex levels and categorise the degree of imposex within 6 assessment classes (A – F; with F representing the highest degree of imposex). Reported trends from the CEMP Assessment Report (2009) present time series data from the coastline of continental Europe (with the majority of study sites along the coastline from Norway, through to Spain, as well as around the UK) (CEMP Assessment Report., 2009). VDSI values have demonstrated a downward trend of imposex levels (from 1997 – 2008), with the majority of imposex categorised within B and C indices levels (which are not considered to affect the reproductive function of gastropods) (CEMP Assessment Report., 2009). Higher imposex values were found within sediments previously identified as contamination hotspots. The reduction observed from the most recent CEMP programme indicates that since the legislative restriction of TBT inputs, a demonstrable recovery of gastropod species within the North Sea region and the UK coastline can be observed (for example, in 2011 the substantial recovery of dog-whelk populations has recently been reported on the south coast of England, between Poole and Selsey) (Matthiessen, 2013).

6. Management and Control of Coastal and Estuarine Sites Contaminated by Organotins in England and Wales

Management strategies within England and Wales for sediments contaminated by organotins fall within statutory legislation controlling dredging activities and the disposal of dredged material at sea. Legacy contaminants within heavily industrialised ports, harbours and estuaries can cause economic and practical problems to those operators having to undertake capital and maintenance dredging. Dredging activities remobilise contaminants within sediments and can cause ecological damage, as well as presenting a potential risk to human health. Dredged material in the United Kingdom is regarded as waste; therefore, disposal is strictly regulated by licensing under section 42 of the Marine and Coastal Access Act, 2009. In cases where contaminated dredged materials are deemed unsuitable for disposal at sea, applicants will need to consider alternate disposal methods. Some of the alternate disposal strategies adopted in the UK and elsewhere are shown in Tables 6 and 7, which are collated from the DEFRA Research and Support for Developing a UK Strategy for Managing Contaminated Marine Sediments (Entec UK limited, 2010).

6.1. Sea Disposal of Dredged Material

Approximately 40 million tons of wet sediment are disposed of annually within dredged material disposal sites around the UK, varying from 26–57 million tons (Bolam et al., 2011). The majority of disposal sites within the UK are located several miles off the coast and are strategically situated within a few miles from major ports or estuary openings, as well as a number located within estuaries themselves (Bolam et al., 2011). England has approximately 150 of these designated sites for the disposal of dredged sediment, however, not all are used in any one year (Bolam et al., 2011). Dredging and subsequent disposal of waste material can have an effect on the water column and the physical and chemical nature of the sea bed (Bolam et al., 2011). This can subsequently have impact to marine ecosystems in both the short- and long-term (Bolam et al., 2006; MEMG, 2003). Distribution of TBT from dredging and disposal operations is of high concern, owing to its high toxicity and sensitivity to a wide range of biota as such both dredging and dredged material disposal are strictly controlled. This is discussed in Section 7.

6.2. Remediation Treatments for TBT Contaminated Sediments and Water

When sea disposal is deemed unsuitable, a number of factors are determined if remediation is to be considered. These include:

1) Degree of contamination by TBT in the water or sediment
2) Volume of contaminated material
3) Nature of TBT contamination
4) Sediment particle sizes
5) Target concentration
6) Availability of the treatment facility
7) Cost

(Adapted from Du et al., 2014)

Remediation techniques capitalise on the high adsorption rate of TBT to particulate matter and its thermal and biological degradation mechanisms. Adsorption methods are typically used as a final step within the waste water remediation procedure by eliminating the lower, dissolved concentrations of TBT (typically for contaminant concentrations at < 1 μg L-1) (Du et al., 2014). Sorbent material used for adsorption of TBT includes modified and natural clays, as well as activated carbon (Vreysen et al., 2008; Du et al., 2014). Aerobic biodegradation techniques of waste water and sediment using microbial communities have proved to be effective, although are slower than other industrial methods (Ayanda et al., 2012). As TBT is inherently toxic to certain microorganisms, microbiological remediation can only be applied to lower levels of sediment contamination (1–10 mg kg^{-1}) (Ayanda et al., 2012; Du et al., 2014). Studies on phytoremediation methods by halophytic plants have reported that sediment colonised with *H. portulacoides* reduced TBT concentrations by 30 % over 12 months (in comparison to non-vegetated sediment) (Carvalho et al., 2010). Thermal treatment of sediments is a well-established technique and is primarily used for the removal of high concentrations of organic compounds from dredged sediments after dewatering. Sediments are placed within a rotating drum and heated to contaminant vaporisation temperature. Volatised compounds are then removed by a cyclone separator and heated > 850°C to aid thermal degradation (Du et al., 2014). Thermal treatment is, however, unsuitable for the removal of heavy metals (Du et al., 2014). Chemical oxidation has received some interest as a viable remediation technique, with reduction of TBT undertaken with by the addition of an oxidation agent (such as potassium permanganate) on mechanically separated fine sediments (Stichnothe et al., 2002). Other chemical treatments include washing with solvents into a waste water stream and have been demonstrated using slops oil, crude oil and diesel oil (Du et al., 2014).

Table collated from information within DEFRA Research and Support for Developing a UK Strategy for Managing Contaminated Marine Sediments (Entec UK limited, 2010).

7. Overview of the Regulations Governing the Management of Dredging and Disposal of Dredged Material in England and Wales

The licensing of dredging and disposal at sea is subject to extensive control. Internationally the London Convention, and its 1996 Protocol, and the OSPAR Convention are treaties concerned with protecting the marine environment from human activities, notably the pollution arising from those activities. The London Convention is a widely applicable treaty which covers the marine waters of the world and focuses on the dumping of waste at sea. The London Protocol was adopted in 1996 to modernise the London Convention, the UK has ratified the London Protocol and so it supersedes the London Convention. The OSPAR Convention is a regional sea convention covering the North East Atlantic region that addresses pollution issues, including those covered by the London Convention/Protocol, at a regional level.

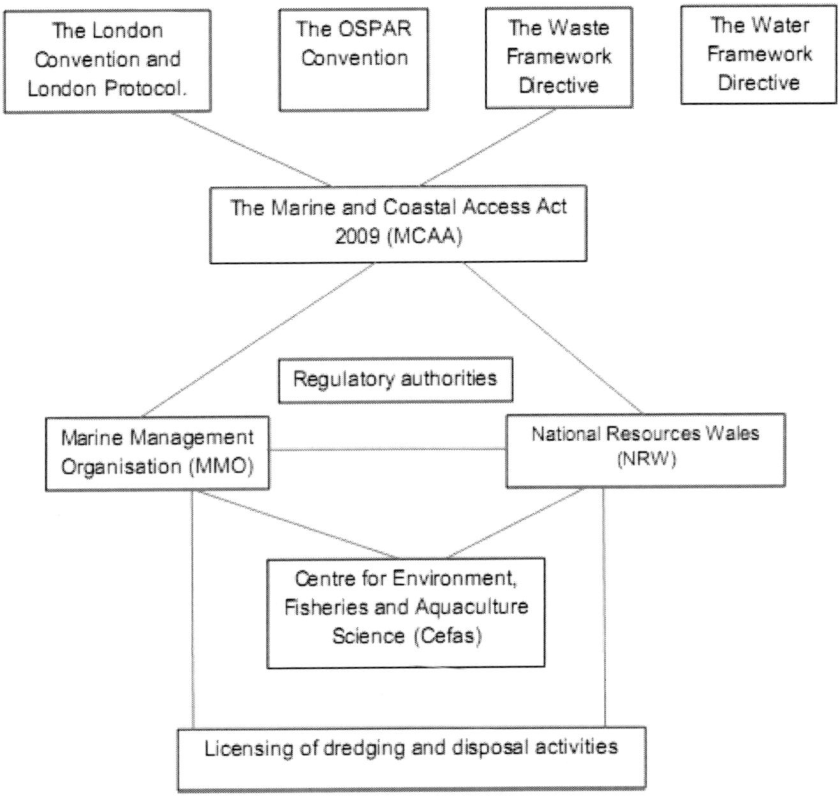

Figure 2. Overview of legislation and regulatory authorites in England and Wales for the licensing of dredging and dredged material disposal activities at sea.

The Water Framework Directive (WFD) is enacted in EU law by Directive 2000/60/EC. The WFD introduced a holistic approach to water management, reflected in a scheme based on river basins requires member states to aim to achieve "good ecological and chemical status" for surface "water bodies" and good qualitative and quantitative status for groundwater bodies by 2015 (unless the criteria for an exemption are met). The WFD is transposed into the law of England and Wales by the Water Environment (Water Framework Directive) (England and Wales) Regulations 2003 (SI 2003/3242) (WFD Regulations). The MMO and NRW are not competent authorities for the purposes of the WFD; however they are required to have regard to the WFD and the WFD Regulations when exercising their marine licensing functions.

The Waste Framework Directive (WaFD) provides the EU legislative framework for the collection, transport, recovery and disposal of waste. The Waste (England and Wales) Regulations 2011 (SI 2011/988) (the Waste Regulations) transpose the WaFD into English law. The Environment Agency is the competent authority responsible for the implementation of the WaFD in England and Wales. However, the MMO and NRW are appropriate authorities under regulation 22(b) of the Waste Regulations and therefore are required to apply the WaFD and, in particular, the waste hierarchy in exercising their 'offshore licensing functions'.

The MCAA provides the principal statutory means by which the UK complies with these international treaties and EU Directives (such the Waste Framework Directive and water Framework Directive) in relation to disposals at sea. Section 71(6) of the MCAA 2009 imposes the legal obligations of both international treaties to the licensing authorities the Marine Management Organisation (MMO) and Natural Resources Wales (NRW) for the regulation and licensing of dredging and dredged material wastes. Licenses for dredging and the disposal of dredged material are issued by the MMO and NRW, under the MCAA 2009, which came into force on the 1st April 2010. The MMO and NRW must ensure that the dredged material is or is not suitable to be disposed to sea or for alternative use. In order for this determination to be made an assessment is undertaken of the physical and chemical characterisation of the dredged material. This process assesses the potential impacts of the disposal of the proposed material to sea and is undertaken in accordance with the OSPAR Guidelines for the Management of Dredged Material (OSPAR 2009).

Cefas provide advice to the MMO and NRW on the suitability of the dredged material for disposal at sea. The chemical analyses undertaken pertain to the contaminant loading of the material and this is assessed against Cefas Action Levels (ALs) and geological background levels. Current Cefas ALs are provided in Table 9 below and are specific to a suite of contaminants broadly grouped under metals, polycyclic aromatic hydrocarbons, polychlorinated biphenyls and organotins.

Concentrations below action level 1 are in general, regarded as low and not enough to effect the licensing decision for sea disposal. Concentrations of contaminants exceeding action level 2 are typically deemed unsuitable for sea disposal, and as such, alternative disposal options would be considered. Further assessments of sediments are undertaken on concentrations that fall between the action levels 1 and 2. Action Levels are used as part of a ‘weight of evidence' approach to assessing dredged material and its suitability for disposal to sea. These values will be used in conjunction with a range of other assessment methods as well as historical data and knowledge regarding the dredging site, the material's physical characteristics, the disposal site characteristics and other relevant data, to make management decisions regarding the fate of dredged material (Bolam et al., 2011). Marine licenses for dredging and disposal at sea are and conditions of compliance are issued by the MMO and NRW to operators.

Table 9. Cefas action levels for dredged material assessment

Contaminant	Action level 1	Action level 2
	mg kg^{-1} dry weight	mg kg^{-1} dry weight
As	20	100
Hg	0.3	3
Cd	0.4	5
Cr	40	400
Cu	40	400
Ni	20	200
Pb	50	500
Zn	130	800
Organotins: TBT DBT MBT	0.1	1
PCB's, sum of ICES 7	0.01	None
PCB's, sum of 25 congeners	0.02	0.2
*DDT	*0.001	
*Dieldrin	*0.005	

*levels set in 1994.

8. Monitoring of TBT at Dredged Material Disposal Sites

The requirement for monitoring of dredging disposal is undertaken as part of the licensing approval process and as such, not all licenses will require monitoring (MEMG, 2003). Licensing conditions relating to disposal monitoring are most likely to be imposed in situations where concern exists over possible dispersal of dredged material to sensitive areas, or where the disposal site itself is identified as potentially sensitive (MEMG, 2003). Cefas conducts an annual programme of disposal site monitoring in England on behalf of the MMO. This programme provides the MMO with supporting information to facilitate licensing decisions, re-evaluate previous decisions and, where necessary, amend licence conditions. (Bolam et al., 2011). Within this monitoring programme, field evaluations of targeted disposal sites are undertaken to assess potential limitations of established sites, change of environmental conditions at new and existing sites and the conformity of disposal licenses at surveyed sites. Cefas use a tier based system to identify sites for monitoring and is based upon the perceived environmental risked posed by the site. Concerns raised by stakeholders are considered in that process (Bolam et al., 2011). Monitoring of these sub-tidal sites are undertaken by sampling sediments from stations situated within and outside the disposal site (Bolam et al., 2014). Samples are analysed for a range of parameters including particle size, contaminants and biota on a site specific basis. Other techniques are also applied in disposal site monitoring, including sediment profile imaging, side scan sonar and multibeam bathymetry (Birchenough et al., 2006) allowing high resolution, *in-situ* assessment of biota and seafloor topography.

The temporal and spatial assessment of TBT from 26 disposal sites along the coast of England and Wales was published in 2014 (Bolam, et al., 2014) analysing data collected and collated from previous Cefas disposal site monitoring programmes. Sampling was undertaken from stations within (IN) and in near proximity (OUT) to the disposal sites. The collection of data used in the study ranges from the year 1998 through to 2010. A selection of 4 sites (Rame Head, Tyne, Souter Point and Tees) for detailed assessment of the concentration TBT was also included for the study. The proportion of samples found to be below the detection limit (below limit of detection (ND) = < 0.002 mg kg^{-1}) were statistically compared across the 26 study sites, with the percentage ND changes plotted at the sampling time intervals. Bolam et al., (2014) reported that the proportion of TBT and DBT concentrations at ND significantly increased over the study period (from around 40% in 1998 to nearly 100% in 2010) with the overall large scale decrease in organotin concentration observed for samples both IN and OUT of the disposal sites. For the 4 selected sites, 3 of the 4 showed a similar decreasing trend for TBT and DBT in concentrations, with one of the sites showing no significant decrease in levels of TBT and DBT. Statistical comparisons of temporal data from IN and OUT sites were undertaken to observe any differences in contamination levels. The percentage of NDs were found to be higher for sites outside of the disposal sites, however from calculating mean concentration changes no significant difference in concentrations within or outside disposal sites was obtained. Results from this recent study show an encouraging reduction of butyltins within disposal sites during the study period. Results are therefore indicative that from the recent material inputs to disposal sites, an overall reduction of total butyltin inputs to English and Welsh disposal sites. It is therefore indicative that current dredged material licensing policy and screening protocols are providing an effective reduction of organotin inputs at dredged material disposal sites.

CONCLUSION

Since the gradual phasing out and prohibition of organotin based antifoulants; recovery from historical organotin contamination within the marine environment has shown encouraging improvements, notably from a reduction of reported sediment and dredged material concentrations and the recorded recovery of TBT bio-indicator species in numerous locations. Current inputs of organotin compounds and their behaviour within marine compartments still remain a concern, notably from their bioaccumulation potential and the subsequent contamination of food stocks. Due to the persistence and very high toxicity of TBT at low concentrations, it is likely that it will remain within statutory regulation for some time. In response, analytical and sampling techniques used to quantify and identify organotins are continually under development to achieve speciation at low limits of detection, in order to meet the higher monitoring demands set by regulators. The research of sampling techniques is also being undertaken to address uncertainties in organotin fluxes within marine environmental compartments. Such avenues of research are now important in understanding the potential risk of remobilisation mechanisms for use in dredged material disposal site monitoring programmes and will be important in deciding future management strategies.

In terms of the increasing economic costs of port management and the requirements of organotin pollutant policy, a further understanding of organotin fluxes within marine compartments will allow regulators to make informed decisions on organotin action levels and licensing parameters, as well as providing further reinforcement with future decisions on best practicable, affordable disposal techniques.

References

Abbott, A., Abel, P. D., Arnold, D. W., Milne, A. (2000). Cost–benefit analysis of the use of TBT: the case for a treatment approach. *Science of The Total Environment, 258* (1–2), 5-19.

Adeloju, S. B. (1991). Adsorptive voltammetric stripping analysis of ultra-trace amounts of tin in natural waters and sediments. *Analytical Sciences, 7* (Supple), 1099-1103.

Aguilar-Martinez, R., Palacios-Corvillo, M., Greenwood, R., Mills, G.A, Vrana, B., Gomez-Gomez, M. (2008). Calibration and use of the Chemcatcher passive sampler for monitoring organotin compounds in water. *Analytica Chimica Acta, 618*(2), 157-167.

Al-rashdi, A. (2011) Improving the measurements of butyltin compounds in environmental samples. Unpublished doctoral thesis, University of Southampton.

Alzieu, C., Sanjuan, J., Deltreil, J. P., Borel, M. (1986). Tin contamination in arcachon bay - effects on oyster shell anomalies. *Marine Pollution Bulletin, 17*(11), 494-498.

Anderson, C., Dalley, R. (1986). Use of Organotins in Antifouling Paints. Paper presented at the OCEANS '86.

Ansari, A. A., Singh, I. B., Tobschall, H. J. (1998). Organotin compounds in surface and pore waters of Ganga Plain in the Kanpur-Unnao industrial region, India. *Science of the Total Environment, 223*(2-3), 157-166.

Antes, F. G., Krupp, E., Flores, E. M. M., Dressler, V. L., Feldmann, J. (2011). Speciation and Degradation of Triphenyltin in Typical Paddy Fields and its Uptake into Rice Plants. *Environmental Science & Technology, 45*(24), Anti-fouling systems. (2014). Retrieved May 11, 2014, from the International Maritime Organisation website.

Antizar-Ladislao, B. (2008). Environmental levels, toxicity and human exposure to tributyltin (TBT)-contaminated marine environment. A review. *Environment International, 34*(2), 292-308.

Arnold, C., Berg, M., Muller, S., Dommann, U., Schwarzenbach, R. (1998). Determination of organotin compounds in water, sediments, and sewage sludge using perdeuterated internal standards, accelerated solvent extraction, and large-volume-injection GC/MS. *Analytical Chemistry, 70*(14), 3094-3101.

Arnold, C., Weidenhaupt, A., David, M., Muller, S., Haderlein, S., Schwarzenbach, R. (1997). Aqueous speciation and 1-octanol-water partitioning of tributyl- and triphenyltin: Effect of pH and ion composition. *Environmental Science & Technology, 31*(9), 2596-2602.

Ayanda, O., Fatoki, O., Asekola, F., Ximba, B. (2012). Fate and Remediation of Organotin Compounds in Seawaters and Soils. *Chemical Science Transactions, 1*(3), 000-000.

Bangkedphol, S., Keenan, H., Davidson, C., Sakultantimetha, A., Songsasen, A. (2009). The partition behavior of tributyltin and prediction of environmental fate, persistence and toxicity in aquatic environments. *Chemosphere, 77*(10), 1326-1332.

Berg, M., Arnold, C. G., Müller, S. R., Mühlemann, J., Schwarzenbach, R. P. (2001). Sorption and Desorption Behavior of Organotin Compounds in Sediment−Pore Water Systems. *Environmental Science & Technology, 35*(15), 3151-3157.

Berto, D., Giani, M., Boscolo, R., Covelli, S., Giovanardi, O., Massironi, M., Grassia, L. (2007). Organotins (TBT and DBT) in water, sediments, and gastropods of the southern Venice lagoon (Italy). *Marine Pollution Bulletin, 55*(10-12), 425-435. doi: 10.1016/j.marpolbul.2007.09.005

Birchenough, S. N. R., Boyd, S. E., Coggan, R. A., Limpenny, D. S., Meadows, W. J., Rees, H. L. (2006). Lights, camera and acoustics: Assessing macrobenthic communities at a dredged material disposal site off the North East coast of the UK. *Journal of Marine Systems, 62*(3–4), 204-216.

Bjorn, A., Horsing, M., Ejlertsson, J., & Svensson, B. H. (2011). Transformation of methyltin chlorides and stannic chloride under simulated landfill conditions. *Waste Management & Research*, 29(12), 1327-1336.

Björn, A., Hörsing, M., Karlsson, A., Mersiowsky, I., Ejlertsson, J. (2007). Impacts of temperature on the leaching of organotin compounds from poly(vinyl chloride) plastics—A study conducted under simulated landfill conditions. *Journal of Vinyl and Additive Technology*, 13(4), 176-188.

Blunden, S. J., Chapman, A. H. (1982). The environmental degradation of organotin compounds — A review. *Environmental Technology Letters, 3*(1-11), 267-272.

Bolam, S. G., Mason, C., Bolam, T., Whomersley, P., Birchenough, S. N. R., Curtis, M., Birchenough, A., Law, R. L. J. (2011). *Dredged Material Disposal Site Monitoring Around the Coast of England: Results of Sampling (2010).* Lowestoft, UK: Centre for Environment, Fisheries and Aquaculture Science (Cefas).

Bolam, S. G., Rees, H. L., Somerfield, P., Smith, R., Clarke, K. R., Warwick, R. M., Garnacho, E. (2006). Ecological consequences of dredged material disposal in the marine environment: A holistic assessment of activities around the England and Wales coastline. *Marine Pollution Bulletin, 52*(4), 415-426.

Bolam, T., Barry, J., Law, R. J., James, D., Thomas, B., Bolam, S. G. (2014). A temporal and spatial assessment of TBT concentrations at dredged material disposal sites around the coast of England and Wales. *Marine Pollution Bulletin, 79*(1-2), 326-332.

Brainina, K., Henze, G., Stojko, N., Malakhova, N., Faller, C. (1999). Thick-film graphite electrodes in stripping voltammetry. *Fresenius' Journal of Analytical Chemistry, 364*(4), 285-295.

Brandli, R. C., Breedveld, G. D., Cornelissen, G. (2009). Tributyltin sorption to marine sedimentary black carbon and to amended activated carbon. *Environmental Toxicology and Chemistry, 28*(3), 503-508.

Briant, N., Bancon-Montigny, C., Elbaz-Poulichet, F., Freydier, R., Delpoux, S., Cossa, D. (2013). Trace elements in the sediments of a large Mediterranean marina (Port Camargue, France): Levels and contamination history. *Marine Pollution Bulletin, 73*(1), 78-85.

Brunori, C., Ipolyi, I., Massanisso, P., Morabito, R. (2006). New Trends in Sample Preparation Methods for the Determination of Organotin Compounds in Marine Matrices. In I. Konstantinou (Ed.), *Antifouling Paint Biocides* (Vol. 50, pp. 51-70): Springer Berlin Heidelberg.

Bryan, G. W., Bright, D. A., Hummerstone, L. G., Burt, G. R. (1993). Uptake, tissue distribution and metabolism of 14C-labelled tributyltin (TBT) in the dog-whelk, Nucella lapillus. *Journal of the Marine Biological Association of the United Kingdom*, 73(04), 889-912.

Buchanan, J. L., McKown, C. (1997). Off-line sheet glass coating system. *Journal of Non-Crystalline Solids, 218*, 179-184.

Burton, E. D., Phillips, I. R., Hawker, D. W. (2004). Sorption and Desorption Behavior of Tributyltin with Natural Sediments. *Environmental Science & Technology, 38*(24), 6694-6700.

Burton, E. D., Phillips, I. R., Hawker, D. W. (2005). In-situ partitioning of butyltin compounds in estuarine sediments. *Chemosphere, 59*(5), 585-592.

Carvalho, P., Basto, M. C., Silva, M. G. M., Machado, A., Bordalo, A. A., Vasconcelos, M. T. D. (2010). Ability of salt marsh plants for TBT remediation in sediments. *Environmental Science and Pollution Research, 17*(6), 1279-1286.

Cassi, R., Tolosa, I., de Mora, S. (2008). A survey of antifoulants in sediments from Ports and Marinas along the French Mediterranean coast. *Marine Pollution Bulletin, 56*(11), 1943-1948.

Centineo, G., Rodriguez-Gonzaqlez, P., & Alonso, I. (2007). Determination of Butyltin Compounds in Environmental Samples by Isotope Dilution GC-MS: Agilent Technologies, Inc.

Centineo, G., Rodriguez-Gonzalez, P., Gonzalez, E., Alonso, J., Sanz-Medel, A. (2004). Simultaneous determination of mono-, di- and tributyltin in environmental samples using isotope dilution gas chromatography mass spectrometry. *Journal of Mass Spectrometry, 39*(5), 485-494.

CEMP assessment report: 2008/2009; Assessment of trends and concentrations of selected hazardous substances in sediments and biota. (2009). Retrieved May 5th 2014, from the OSPAR Co-ordinated Environmental Monitoring Programme website: http://www.ospar.org/content/content.asp

Cima, F., Craig, P., Harrington, C. (2003). *Organometallic Compounds in the Environment* (Second ed.). De Montfort University, Leicester, UK: John Wiley & Sons Ltd.

Díez, S., Ábalos, M., Bayona, J. M. (2002). Organotin contamination in sediments from the Western Mediterranean enclosures following 10 years of TBT regulation. *Water Research, 36*(4), 905-918.

dos Santos, D. M., Sant'Anna, B. S., Sandron, D. C., de Souza, S. C., Cristale, J., de Marchi, M. R. R., Turra, A. (2010). Occurrence and behavior of butyltins in intertidal and shallow subtidal surface sediments of an estuarine beach under different sampling conditions. *Estuarine Coastal and Shelf Science, 88*(3), 322-328.

Du, J., Chadalavada, S., Chen, Z., Naidu, R. (2014). Environmental remediation techniques of tributyltin contamination in soil and water: A review. *Chemical Engineering Journal, 235*(0), 141-150.

Dubey, S. K., Roy, U. (2003). Biodegradation of tributyltins (organotins) by marine bacteria. *Applied Organometallic Chemistry, 17*(1), 3-8.

Entec UK Limited. (2010). Research and Support for Developing a UK Strategy for Managing Contaminated Marine Sediments; Task 5: Establishing Best Practise for Current Disposal Treatment Options for Contaminated Dredged Marine Sediments. Newcastle upon Tyne. Entec UK Limited.

Epstein, R. L., Phillippo, E. T., Harr, R., Koscinski, W., Vasco, G. (1991). Organotin residue determination in poultry and turkey sample survey in the united-states. *Journal of Agricultural and Food Chemistry, 39*(5), 917-921.

Eschenbach. A, Höhne. J, Söö. Z, Luther. G The steam stripping process: a remediation technique for TBT- and PAH-contaminated dredged sediments and soils: Proceedings of 2001 International Containment and Remediation Technology Conference and Exhibition, Orlando, FL, 2001, p. 3.

Fargasova, A., Reinprecht, L., Kizlink, J. (1997). Efficiency of organotin dithiocarbamate derivatives against wood destroying fungi. *Biologia, 52*(3), 451-455.

Ferri, T., Roberti, F., Chiavarini, S., Cremisini, C., Morabito, R. (1994). Influence of tropolone on voltammetric speciation analysis of butyltin compounds. *Applied Organometallic Chemistry, 8*(7-8), 587-593.

Filipkowska, A., Kowalewska, G., Pavoni, B. (2014). Organotin compounds in surface sediments of the Southern Baltic coastal zone: a study on the main factors for their accumulation and degradation. *Environmental Science and Pollution Research, 21*(3), 2077-2087.

Filipkowska, A., Kowalewska, G., Pavoni, B., Łeczynski, L (2011). Organotin compounds in surface sediments from seaports on the Gulf of Gdańsk (southern Baltic coast). *Environmental Monitoring & Assessment, 182*(1-4), 455-466.

Flores, M., Bravo, M., Pinochet, H., Maxwell, P., Mester, Z. (2011). Tartaric acid extraction of organotin compounds from sediment samples. *Microchemical Journal, 98*(1), 129-134.

Folsvik, N., Brevik, E. M., Berge, J. A. (2000). Monitoring of organotin compounds in seawater using semipermeable membrane devices (SPMDs)-tentative results. *Journal of Environmental Monitoring, 2*(4), 281-284.

Gadd, G. M. (2000). Microbial interactions with tributyltin compounds: detoxification, accumulation, and environmental fate. *Science of The Total Environment, 258*(1–2), 119-127.

Galante-Oliveira, S., Oliveira, I., Santos, J. A., Pereira Mde, L., Pacheco, M., Barroso, C. M. (2010). Factors affecting RPSI in imposex monitoring studies using Nucella lapillus (L.) as bioindicator. *J. Environ. Monit.*, 12(5), 1055-1063.

Gao, Z., Siow, K.S. (1996). Adsorptive Stripping Differential Pulse Voltammetric Determination of Trace Amounts of Tin in Biological Samples. *Analytical Sciences, 12*(2), 267-271.

Gibbs, P. E., Bryan, G. W. (1986). Reproductive failure in populations of the dog-whelk, nucella-lapillus, caused by imposex induced by tributyltin from antifouling paints. Journal of the Marine Biological Association of the United Kingdom, 66(4), 767-&.

Giltrap, M., Macken, A., McHugh, B., McGovern, E., Foley, B., Davoren, M. (2011). In vitro screening of organotin compounds and sediment extracts for cytotoxicity to fish cells. *Environmental Toxicology and Chemistry, 30*(1), 154-161.

Gomez-Ariza, J. L., Giraldez, I., Morales, E., Ariese, F., Cofino, W., Quevauviller, P. (1999). Stability and storage problems in organotin speciation in environmental samples. *Journal of Environmental Monitoring, 1*(2), 197-202.

González-Toledo, E., Compañó, R., Granados, M., Dolors Prat, M. (2003). Detection techniques in speciation analysis of organotin compounds by liquid chromatography. *TrAC Trends in Analytical Chemistry, 22*(1), 26-33.

Graceli, J. B., Sena, G. C., Lopes, P. F. I., Zamprogno, G. C., da Costa, M. B., Godoi, A. F. L., dos Santos, D.M., de Marchi, M.R.R., dos santos Fernandez, M. A. (2013). Organotins: A review of their reproductive toxicity, biochemistry, and environmental fate. *Reproductive Toxicology, 36*(0), 40-52.

Harino, H., Arifin, Z., Rumengan, I. F. M., Arai, T., Ohji, M., Miyazaki, N. (2012). Distribution of Antifouling Biocides and Perfluoroalkyl Compounds in Sediments From Selected Locations in Indonesian Coastal Waters. *Archives of Environmental Contamination and Toxicology, 63*(1), 13-21. doi: 10.1007/s00244-011-9747-y.

Harman, C., Boyum, O., Tollefsen, K. E., Thomas, K., Grung, M. (2008). Uptake of some selected aquatic pollutants in semipermeable membrane devices (SPMDs) and the polar organic chemical integrative sampler (POCIS). *Journal of Environmental Monitoring, 10*(2), 239-247.

Harrington, C. F., Eigendorf, G. K., Cullen, W. R. (1996). The use of High-Performance Liquid Chromatography for the Speciation of Organotin Compounds. *Applied Organometallic Chemistry, 10*(5), 339-362.

Heppeler, F., Sander, S., Henze, G. (1996). Determination of tin traces in water samples by adsorptive stripping voltammetry. *Analytica Chimica Acta, 319*(1–2), 19-24.

Hoch, M. (2001). Organotin compounds in the environment - an overview. *Applied Geochemistry, 16*(7-8), 719-743.

Hoch, M., Alonso-Azcarate, J., Lischick, M. (2002). Adsorption behavior of toxic tributyltin to clay-rich sediments under various environmental conditions. *Environmental Toxicology and Chemistry, 21*(7), 1390-1397.

Horiguchi, T., Ohta, Y., Urushitani, H., Lee, J. H., Park, J. C., Cho, H. S., & Shiraishi, H. (2012). Vas deferens and penis development in the imposex-exhibiting female rock shell, Thais clavigera. *Mar. Environ. Res,* 76, 71-79.

Huang, G. L., Chen, C. J., & Dai, S. G. (1998). Transport dynamics of tributyltin between the surface microlayer and subsurface water. *Applied Organometallic Chemistry*, 12(8-9), 559-563.

IUPAC. (2011). Fentin chloride (Ref: Hoe 2872). Pesticides Property database. Retrieved 20th March, 2013.

Jadhav, S., Bhosle, N. B., Massanisso, P., Morabito, R. (2009). Organotins in the sediments of the Zuari estuary, west coast of India. *Journal of Environmental Management, 90*, S4-S7.

Jiang, G. B., Xu, F. Z. (1996). Speciation analysis of butyltin species in water by gas chromatography with flame photometric detection using quartz surface-induced tin emission. *Applied Organometallic Chemistry, 10*(2), 77-82.

Junzo, Y., Hiroaki, S., & Yuko, S. (1993). In vitro assessment of teratogenic potential of organotin compounds using rat embryo limb bud cell cultures. *Toxicology Letters*, 66(2), 183-191.

Kim, N. S., Shim, W. J., Yim, U. H., Hong, S. H., Ha, S. Y., Han, G. M., Shin, K. H. (2014). Assessment of TBT and organic booster biocide contamination in seawater from coastal areas of South Korea. *Marine Pollution Bulletin, 78*(1-2), 201-208.

Kurihara, R. Y. O., Rajendran, R. B., Tao, H., Yamamoto, I., Hashimoto, S. (2007). Analysis of organotins in seawater of the southern ocean and suruga bay, japan, by gas chromatography/inductively coupled plasma mass spectrometry. *Environmental Toxicology & Chemistry, 26*(4), 647-654.

Labare, M. L., Coon, S. L., Matthias, C., Weiner, R. M. (1997). Magnification of tributyl tin toxicity to oyster larvae by bioconcentration in biofilms of Shewanella colwelliana. *Appl. Environ. Microbiol.*, 63(10), 4107-4110.

Leermakers, M., Nuyttens, J., Baeyens, W. (2005). Organotin analysis by gas chromatography-pulsed flame-photometric detection (GC-PFPD). *Anal. Bioanal. Chem.*, 381(6), 1272-1280.

Leroy, M. J. F., Quevauviller, P., Donard, O. F. X., Astruc, M. (1998). Determination of tin species in environmental samples (Technical Report). *Pure and Applied Chemistry, 70*(10), 2051-2064.

Liu, L.L., Wang, J.T., Chung, K.N., Leu, M.Y., Meng, P.J. (2011). Distribution and accumulation of organotin species in seawater, sediments and organisms collected from a Taiwan mariculture area. *Marine Pollution Bulletin, 63*(5–12), 535-540.

Looser, P. W., Bertschi, S., Fent, K. (1998). Bioconcentration and bioavailability of organotin compounds: Influence of pH and humic substances. *Applied Organometallic Chemistry, 12*(8-9), 601-611.

Ma, H., Dai, S., Huang, G. (2000). Distribution of tributyltin chloride in laboratory simulated estuarine microcosms. *Water Research, 34*(10), 2829-2841.

Marine licensing. (2014) Retrieved May 11, 2014, from the Marine Management Organisation Website.

Marine Maritime Organisation. (2014). Licensable activities and exemptions. Retrieved 22nd April, 2014, from http://www.marinemanagement.org.uk /licensing/marine/activities.htm

Martin, F. M., Tseng, C. M., Belin, C., Quevauviller, P., Donard, O. F. X. (1994). Interferences generated by organic and inorganic-compounds during organotin speciation using hydride generation coupled with cryogenic trapping, gas-chromatographic separation and detection by atomic-absorption spectrometry. *Analytica Chimica Acta, 286*(3), 343-355.

Matthiessen, P. (2013). Detection, monitoring, and control of tributyltin-an almost complete success story. *Environmental Toxicology & Chemistry,* 32(3), 487-489.

MEMG. (2003). *Group Co-ordinating Sea Disposal Monitoring. Final Report of the Dredging and Dredged Material Disposal Monitoring Task Team Sci. Ser., Aquat. Environ. Monit. Rep.,*. Lowestoft: Cefas.

Mizuishi, K., Takeuchi, M., Hobo, T. (1998). Effect of hydrogen bromide doping on capillary gas chromatographic analysis of tributyltin and triphenyltin halides. *Analyst, 123*(2), 329-335.

Morabito, R., Massanisso, P., Quevauviller, P. (2000). Derivatization methods for the determination of organotin compounds in environmental samples. *Trac-Trends in Analytical Chemistry, 19*(2-3), 113-119.

Negri, A. P., Heyward, A. J. (2001). Inhibition of coral fertilisation and larval metamorphosis by tributyltin and copper. *Marine Environmental Research*, 51(1), 17-27.

Nyberg. E, Poikane. R, Strand. J, Larsen. M.M, Danielsson. S, Bignert, A. (2013). Tributyltin (TBT) and imposex, HELCOM Core Indicator Report Online. Retrieved 27th May 2014, http://helcom.fi/Core%20Indicators/ HELCOM_CoreIndicator_Tributyltin_and_imposex.pdf

Okoro, H., Fatoki, O., Adekola, F., Ximba, B., Snyman, R. (2010). Sources, Environmental Levels and Toxicity of Organotin in Marine Environment - A Review. *Asian Journal of Chemistry, 23*(2), 473 - 482.

Oliveira, R., Santelli, R. (2010). Occurrence and chemical speciation analysis of organotin compounds in the environment: A review. *Talanta, 82*(1), 9-24.

Omae, I. (2003). Organotin antifouling paints and their alternatives. *Applied Organometallic Chemistry, 17*(2), 81-105.

OSPAR Commision. (2009). *Assessment of Impacts of shipping on the marine environment*. London, United Kingdom: Ospar Commision.

OSPAR Commision. (2013). *JAMP Guidelines for Monitoring Contaminants in Sediments (agreement 2006-16)*. London, UK: OSPAR.

Paton, G. I., Cheewasedtham, W., Marr, I. L., Dawson, J. J. C. (2006). Degradation and toxicity of phenyltin compounds in soil. *Environmental Pollution, 144*(3), 746-751.

Quevauviller, P., Donard, O. F. X., Bruchet, A. (1991). Leaching of organotin compounds from poly(vinyl chloride) (PVC) material. *Applied Organometallic Chemistry, 5*(2), 125-129.

Radke, B., Staniszewska, M., Wasik, A., Namieśnik, J., Bolałek, J. (2008). Organotin Compounds in Marine Sediments. *Polish Journal of Environmental Studies, 17*(5), 643-654.

Radke, B., Wasik, A., Jewell, L. L., Piketh, S., Paczek, U., Galuszka, A., Namiesnik, J. (2012). Seasonal changes in organotin compounds in water and sediment samples from the semi-closed Port of Gdynia. *Science of the Total Environment, 441*, 57-66.

Readman, J. W., Kwong, L. L. W., Grondin, D., Bartocci, J., Villeneuve, J. P., & Mee, L. D. (1993). Coastal water contamination from a triazine herbicide used in antifouling paints. *Environmental Science & Technology*, 27(9), 1940-1942.

Rodríguez, J. G., Solaun, O., Larreta, J., Belzunce Segarra, M. J., Franco, J., Ignacio García Alonso, J., Borja, Á. (2010). Baseline of butyltin pollution in coastal sediments within the Basque Country (northern Spain), in 2007–2008. *Marine Pollution Bulletin, 60*(1), 139-145.

Rodríguez-González, P., Encinar, J., Alonso, J., Sanz-Medel, A. (2006). Contamination of the Coastal Waters of Gijón (North West Spain) by Butyltin Compounds. *Water, Air & Soil Pollution, 174*(1-4), 127-139.

Sadiki, A.D., Williams, D. T. (1999). A study on organotin levels in Canadian drinking water distributed through PVC pipes. *Chemosphere, 38*(7), 1541-1548.

Seligman, P. F., Maguire, R. J., Lee, R. F., Hinga, K. R., Valkirs, A. O., Stang, P. M. (1996). Persistence and Fate of Tributyltin in Aquatic Ecosystems. In M. Champ & P. Seligman (Eds.), *Organotin* (pp. 429-457): Springer Netherlands.

Shreadah, M. A., Said, T. O., Abd El Ghani, S. A., M. Ahmed, A. E. M. (2011). Distribution of Different Organotin and Organolead Compounds in Sediment of Suez Gulf. *Journal of Environmental Protection, 2*(5), 545-554.

Smedes, F., Beeltje, H. (2011). Passive Sampling of organotin compounds (pp. 38). Utrecht, The Netherlands: Deltares Geo-environmental Research Laboratory.

Smedes, F., de Jong, A., Davies, I. (2000). Determination of (mono-, di- and) tributyltin in sediments. Analytical methods. *Journal of Environmental Monitoring, 2*(6), 541-549.

Stichnothe, H., Keller, A., Thoming, J., Lohmann, N., Calmano, W. (2002). Reduction of tributyltin (TBT) and other organic pollutants of concern in contaminated Sediments by means of an electrochemical oxidation. *Acta Hydrochimica Et Hydrobiologica, 30*(2-3), 87-93.

Tanguy, A., Fernandez Castro, N., Marhic, A., & Moraga, D. (1999). Effects of an Organic Pollutant (Tributyltin) on Genetic Structure in the Pacific Oyster *Crassostrea gigas. Marine Pollution Bulletin, 38(7), 550-559.*

The World Health Organisation. (1990). Tributyltin Compounds. Environmental Health Criteria 116.International Programme on Chemical Safety. World Health Organization, Geneva, Switzerland.

The World Health Organisation. (1999). Triphenyltin Compounds. In WHO (Ed.), (pp. 40). Geneva,Switzerland.

The World Health Organisation. (2006). *Mono- and disubstituted methyltin, butyltin, and octyltin compounds.* Geneva, Switzerland: The World Health Organisation.

Tsunoi, S., Matoba, T., Shioji, H., Huong Giang, L. T., Harino, H., Tanaka, M. (2002). Analysis of organotin compounds by grignard derivatization and gas chromatography–ion trap tandem mass spectrometry. *Journal of Chromatography A*, 962(1–2), 197-206.

Unger, M. A., Huggett, R. J., MacIntyre, W. G. (1996). Sorption Behavior of Tributyltin. In M. Champ & P. Seligman (Eds.), *Organotin* (pp. 475-483): Springer Netherlands.

van den Berg, C. M. G., Khan, S. H., Riley, J. P. (1989). Determination of tin in sea water by adsorptive cathodic stripping voltammetry. *Analytica*

Chimica Acta, 222(1), 43-54. doi: http://dx.doi.org/10.1016/S0003-2670(00)81878-4.

Vella, A. J., Adami, J. P. T. (2001). Preliminary evidence for in vitro methylation of tributyltin in a marine sediment. *Applied Organometallic Chemistry, 15*(11), 901-906.

Vella, A. J., & Vassallo, R. (2002). Emission to air of volatile organotins from tributyltin contaminated harbour sediments. *Applied Organometallic Chemistry, 16*(5), 239-24

Vreysen, S., Maes, A., Wullaert, H. (2008). Removal of organotin compounds, Cu and Zn from shipyard wastewaters by adsorption-flocculation: a technical and economical analysis. *Mar. Pollut. Bull, 56*(1), 106-115.

Wahlen, R. (2002). A comparison of GC-ICP-MS and HPLC-ICP-MS for the analysis of organotin compounds. *Lc Gc Europe, 15*(10), 670-+.

Waldock, M. J., Thain, J. E. (1983). Shell thickening in *crassostrea gigas* - organotin anti-fouling or sediment induced? *Marine Pollution Bulletin, 14*(11), 411-415.

Wang, J., Larson, D., Foster, N., Armalis, S., Lu, J., Rongrong, X., Zirino, A. (1995). Remote Electrochemical Sensor for Trace Metal Contaminants. *Analytical Chemistry, 67*(8), 1481-1485.

Wang, X., Hong, H., Zhao, D., Hong, L. (2008). Environmental behavior of organotin compounds in the coastal environment of Xiamen, China. *Marine Pollution Bulletin, 57*, 419-424.

Weber, G. (1986). Determination of tin in the ng g−1 range differential pulse polarography. *Analytica Chimica Acta, 186*(0), 49-56.

Wetzel, M. A., Wahrendorf, D. S., von der Ohe, P. C. (2013). Sediment pollution in the Elbe estuary and its potential toxicity at different trophic levels. *Science of the Total Environment, 449*, 199-207.

Wetzel, M. A., Winterscheid, A., Wahrendorf, D. S. (2013). Baseline of the butyltin distribution in surface sediments (0-20 cm) of the Elbe estuary (Germany, 2011). *Marine Pollution Bulletin, 77*(1-2), 418-423.

World Health Organisation. (1990). *Tributyltin Compounds.* Geneva, Switzerland: World Health Organisation.

Yamada, S., Fujii, Y., Mikami, E., Kawamura, N., Hayakawa, J., Aoki, K., Terao, C. (1993). Small-scale survey of organotin compounds in household commodities. *Journal of Aoac International, 76*(2), 436-441.

Yebra, D. M., Kiil, S., & Dam-Johansen, K. (2004). Antifouling technology - past, present and future steps towards efficient and environmentally friendly antifouling coatings. *Progress in Organic Coatings*, 50(2), 75-104.

Zhang, K. G., Shi, J. B., He, B., Xu, W. H., Li, X. D., Jiang, G. B. (2013). Organotin compounds in surface sediments from selected fishing ports along the Chinese coast. *Chinese Science Bulletin, 58*(2), 231-237.

Zuo, Z., Chen, S., Wu, T., Zhang, J., Su, Y., Chen, Y., & Wang, C. (2011). Tributyltin causes obesity and hepatic steatosis in male mice. *Environ. Toxicol., 26*(1), 79-85.

In: Marine Pollution
Editor: Dominic E. Jefferson
ISBN: 978-1-63321-520-7

Chapter 2

MANAGEMENT OF LARGE OIL SPILLS

A. García-Olivares, E. García-Ladona and J. A. Jiménez Madrid
Institute of Marine Sciences (ICM)
Spanish Council for Scientific Research (CSIC) Barcelona, Spain

ABSTRACT

Large oil spills produce major ecological and economic impacts on shorelines depending on the following conditions: (i) persistence of the oil in the environment; (ii) weather and sea conditions; (iii) amount and rate of spillage; (iv) geographical, biological and economic characteristics of the area affected; (v) effectiveness of clean up. Oil spill dispersion models are necessary tools to forecast the trajectory and persistence of oil slicks following a marine accident. However, their reliability depends on the quality of available meteorological and hydrological data and assimilation strategies. Large oil spills that took place in regions relatively distant from the coast, such as the Prestige accident in November 2002 and the Deep Water Horizon explosion in April 2010, showed the need for having both a short-term forecast for the days following the accident as well as a long-term forecast to prevent future pollution impact. However, there is still no universally accepted methodology for the management of this kind of oil spill. A combined forecast of the short-term deterministic trajectory and the long-term probabilistic dispersion of the slicks may be useful to delimit the most probable consequences of any spill. Medess-4MS is an example of an integrated operational multi-model oil spill system which is under development to meet both requirements. This double methodology is illustrated by the analysis of the oil spill produced by the supertanker Prestige near the Spanish coast in November 2002. This double forecast makes it possible to obtain the short-term and long-term probability of oil beaching on different shoreline segments, followed by preliminary estimates of ecological risk, and economic cost based on the analysis of past similar accidents.

INTRODUCTION

Pollution enters the marine environment in three ways. The first route is from land to sea, as in the case of waste disposal in continental waters in areas of human agglomerations, through rivers and estuaries or during events associated to accidents of industrial factories and during severe floods in coastal areas. The second way is through direct pollution associated with global maritime trade activity (Fig. 1) of ships where noxious substances are directly released in ship accidents, from illegal or unregulated activities. They then are released from industrial platforms at sea, typically oil-gas platforms. A third way, fortunately less recurrent, is from atmospheric deposition of substances as they occur with radioactive clouds (i.e. Fuckushima nuclear waste, Buesseler et al., 2011).

Oil spill pollution events belong to the first two categories. While big spills are often associated with oil tanker accidents (i.e. Exxon Valdez, Prestige) there are many examples of land to sea pollution events generally associated with war conflicts (War of Gulf in 1991, Lebanon spill in 2006) and direct release from oil-gas platforms with non-negligible quantities spilled (i.e. Deep Water Horizon in 2010). Nevertheless, the major number of events of large oil spills are associated with ship accidents, mostly damaged oil tankers but also fuel leaking from non-tanker ship incidents. According to the statistics of the International Tanker Owners Pollution Federation Limited (ITOPF), the cumulative data on oil spills from oil tankers in the last forty years show a decreasing tendency of the total number of oil spills (Fig. 2) and similarly of the total amount of oil spilt (see Figure 6 in ITOPF, 2013).

This evolution is in part explained by an improvement of ship design and navigation security but also as a consequence of more severe regulations adopted through international agreements (see e.g. MARPOL, 1978; OPA, 1990; ISM-Code, 1993). In the 1990s there were around 360 spills over 7 tons, resulting in 1,137,000 tons of oil lost; 73% of this amount was spilled in just 10 incidents. In the 2000s, there were 182 spills over 7 tons, almost half compared to the preceding decade, resulting in 212,000 t of oil lost; 47% of this amount was spilled in just 2 incidents.

To have an idea of the regions that are most exposed to suffer incidents involving an oil spill, Figure 4 represents the cumulative number of ITOPF attendance to oil tanker incidents since 1970. It can be observed that the European and Asiatic coastal regions particularly stand out from the rest. The Mediterranean basin is another example of such a concentration of incidents. It supports an important amount of oil tanker traffic (Fig. 4) because is the shortest route from Asia to Europe. About 1/6 of the global sea traffic and 1/3 of the global oil traffic, corresponding to almost 8 million oil barrels per day, is carried through the Mediterranean basin which represents only 0.8% of the ocean surface. Most oil tankers use the Strait of Gibraltar and the Suez Channel as the main route for traffic between northern Europe and the Middle East and through the Bosphorus Strait to the Black Sea terminals. In a report by the Regional Marine Pollution Emergency Response Centre for the Mediterranean Sea (REMPEC, 2011) there has been a total spillage of around 310.000 tons of oil since 1977 and at least 120.000 tons of other noxious substances since 1988. There were a total of 659 accidents with 545 involving oil substances (Fig. 5). The updated data, available since 1977, also shows a similar decreasing trend since the 90's both for the quantities of oil and other hazardous and noxious substances, only disrupted by the major accidents of the MT-HAVEN and Lebanon spills. The HAVEN accident has been by far the largest in the Mediterranean not only because the quantity spilled was quite huge (about 145.000 T, among the top ten biggest spills from tankers) but also because of the environmental damage. The accident caused serious pollution in the sea waters, the seabed and along the Ligurian coast from Genoa to Savona, producing a shipwreck that still today constitutes a potential source of pollution.

The amount released in the MT-HAVEN incident was comparable, and even greater than other devastating accidents around the world (i.e. Deep Water Horizon, Prestige, Erika, Exxon Valdez). The environmental impact was relatively less because the circumstances were quite distinct. The proximity to the coast, the rapid response, and the meteorological and oceanographic conditions were elements that limited the damage. Thus, the environmental impact by oil spills, or in general by any marine pollution event, is a complex combination of several individual factors and circumstances that make it difficult to totally prevent it.

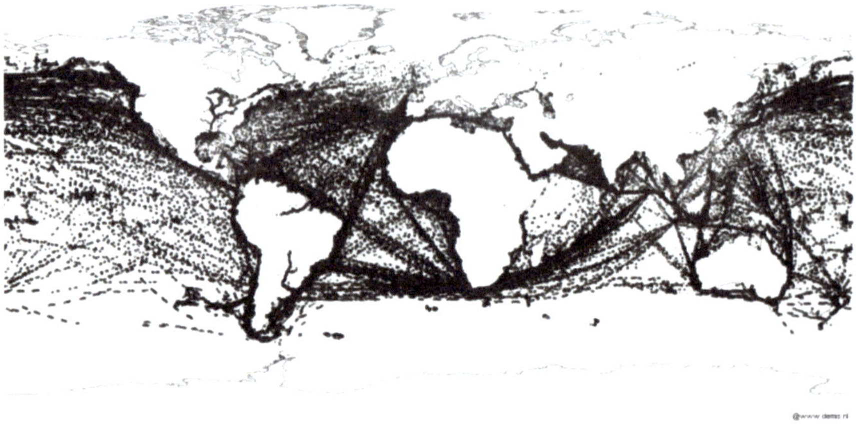

Figure 1. Cumulative positions of ship paths collected between 2-7 June, 2011 from AIS system.(© ESA, www.dennis.nl).

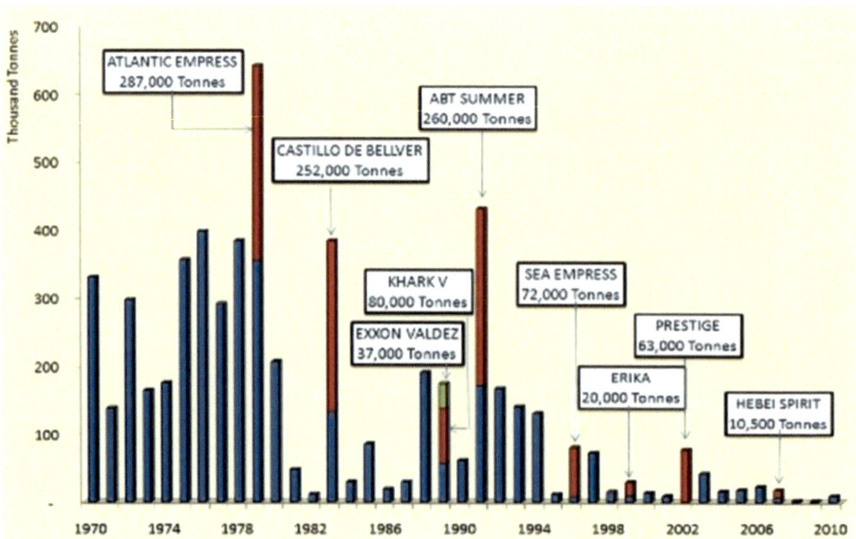

Figure 2. Quantity of oil spilt from oil tankers since 1970 according to ITOPF (reproduced with ITOPF permission).

As in many other situations, being at no risk at all for such types of accidents is almost impossible, and even with strict regulations, safety rules and recommendations, an accident can occur at any time. An example is the Deep Water Horizon oil drilling rig which, although not accounted for in the previous statistics, has produced the biggest spill of around 4.9 million oil barrels (BP Deepwater Horizon Oil Budget: What Happened to the Oil?, US geological Survey Freedom of Information Act, 2010, http://astro.berkeley.edu /~echiang/bp/WorkingPaper.pdf).

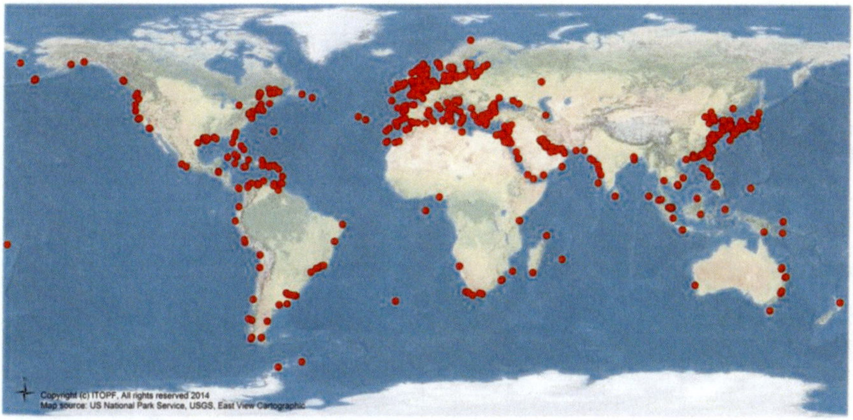

Figure 3. Spills attended by ITOPF staff, 1970-2013 (reproduced with ITOPF permission).

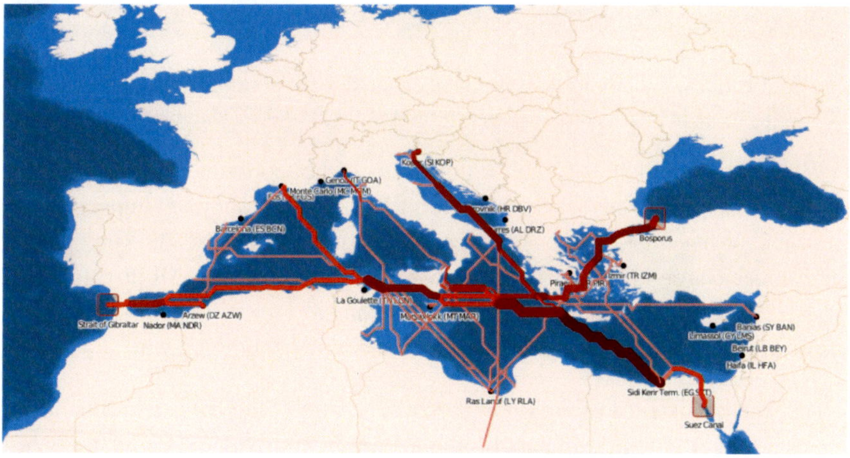

Figure 4. Main paths of oil tankers through the Mediterranean basin. Black and grey dots are main ports where crude oil tankers have been calling in 2005. Red lines are the main routes for crude oil tankers for 2005 and the orange lines are the main traffic routes for the same year. (Source: Med GIS on Maritime Traffic from EU funded Safemed Project, see http://www.safemedgis.org/).

In a marine context, prevention measures are indispensable but systems to efficiently react in front of emergency situations are also crucial to manage and mitigate their impact. Nowadays according to the data, the probability of a big accident is decreasing and the regulations are effective in lowering the risk levels, but regulations and agreements often appear as a response to new incidents, which distort the previous statistics (e.g. the Oil Pollution Act of 1990, was largely driven by the catastrophic EXXON VALDEZ spill, Burns et al. 2002).

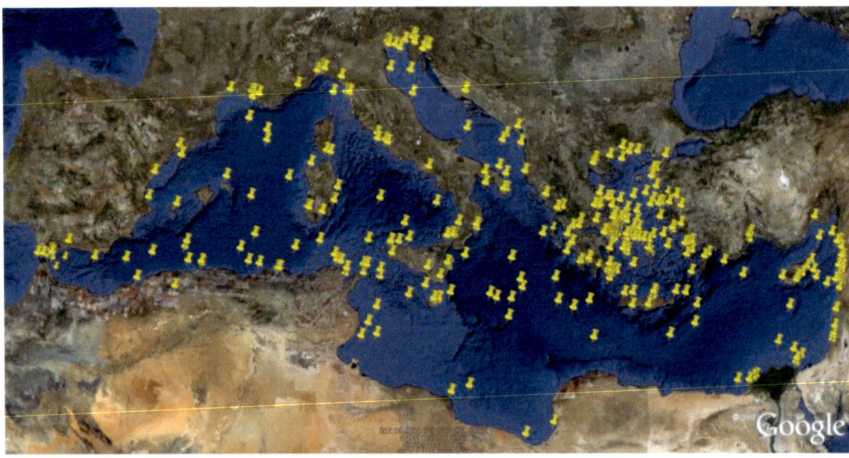

Figure 5. Locations of all the marine accidents related with oil spills in the Mediterranean Basin from 1991 to 2010 reported in the REMPEC database.

When one incident of that kind occurs, the consequences linked to environmental pollution, cleaning operations, and economic impacts are enormous. What makes the differences among them are the particular circumstances around each event. According to White and Molloy (2003) there is a general agreement on the main technical factors that influence the costs of spills. These factors can be enumerated in the following way.

1. The persistence of the oil in the environment; long persistent oils have a longer potential impact on the economy, tourism and ecosystems of the affected coast.
2. Weather and sea conditions; the dominant winds and currents may lead the main fraction of the spill into the coast or out to the deep sea, resulting in different impacts. In this regard, the season of the year may be important depending on the geographical location.

3. Amount spilled and rate of spillage.
4. Geographical, biological and economic characteristics of the area affected.
5. Effectiveness of clean-up. A quick intervention by a fleet of cleaner ships may be able to confine the spill around the ship, minimizing the long-term impacts. A clean-up of an already contaminated coast may have very different short-term and long-term costs.

In case of an oil spill incident, the most pressing issue is to find out if the oil slick will move into a coastal area in the days following the spill. Even for state-of-the-art oil spill forecasting, the prediction horizon is still about 5-7 days because of the prediction horizon imposed by meteorological forecasting. If the 7-day forecast predicts that the slick still will not have reached any shoreline, the coastal segments which are the most exposed to the long-term drift of the slick remain uncertain. In this situation, the initial-condition forecast does not facilitate a fast response and the decision maker may perceive the use of forecasts on initial conditions as useless.

This can be solved by including statistical information of the coastal segments that are most exposed (in the scale of months) to spills coming from the most probable potential sources, such as the marine positions that are most frequently visited by tankers, the oil-gas platforms, etc. In this way, two kind of relevant information can be offered to a decision maker: (i) forecasts of where the spill will be drifting during the following days (up to 7 days), and (ii) long-term (more than 7 days) probability of beaching into different coastal shorelines. Both are relevant for making the first decisions during emergency situations. The short-term can be solved through combined meteorological-oceanographic forecasting systems coupled with an oil spill forecast model, while the long-term can be based on the statistical analysis of past environmental wind fields and marine velocities.

In this chapter we will review the existing strategies and requirements to satisfy both aspects. In Section 2 we will present a brief overview on actions taken in some past accidents to analyze the limitations and drawbacks in the management during the spills events. Section 3 will discuss the bases of state-of-the-art of oil spill forecasting systems commonly in use. Focus will be put on the oil spills modeling part and on the observational strategies to obtain the best first guess of initial conditions possible, both are key elements to improve the accuracy of the forecasts. Section 4 discusses the two modeling strategies that are presented in this work to assess decision-makers facing an oil spill emergency. In Section 5, we illustrate the methodology proposed, with a case study based on a reanalysis of the Prestige accident.

2. Some Past Incidents

2.1. The "Prestige" Incident

On the 13th of November 2002, under very bad weather conditions, the tanker Prestige, en route to Gibraltar, sustained hull damage 15 nm (nautical miles) away from the Galician coast.

The Spanish authorities ordered to tug the tanker away from the coast taking a route, first northwest and then southwest, until it collapsed in two parts sinking on the 19th of November at 42°12' N and 12°03' W over the area called the Galicia's bank (3500 m depth, *Informe sobre el accidente del Buque "Prestige"*, 2002; CEDRE 2002). The tanker continuously leaked crude oil along its trajectory and produced a large oil spill when it broke, releasing a total estimated amount of 66.000 tonnes of crude oil and producing what has been considered the most catastrophic pollution event along the Spanish coast. Three main oil release events were reported from the Prestige tanker: the first one was a mobile, continuous release from the vessel's initial position on the 13th of November to the sinking position on the 19th of November, which amounted 6,000 tonnes; the second one took place in the sinking area and released an uncertain quantity larger than 20,000 tonnes; A third release was reported from the sunken wreckage until the blocking of the leakages in the hull (January 29, 2003) accounting for about 4,000 tonnes (Montero et al. 2003).

According to the official sources more than 1000 km of coast was affected including some parts of the French coast in the Gulf of Biscay. Figure 4 shows the well-known satellite image which illustrates and synthesizes the Prestige incident. The thickest slick connecting the sinking point and the coast produced the first big spill reaching the coast at the area around Cape Finisterre.

The black regions in the estuary areas south of the Galician coast do not correspond to the oil spill but the low wind areas due to the screening effect of the local topography.

The characteristics of the incident make it very particular in terms of oceanographic conditions with respect to other past events. The release was important and produced in a continuous way, initially over the shelf, then relatively far from the coast and finally in the open ocean. Second, at that time

Spain did not have an operational system that could provide the necessary forecast to assess cleaning and mitigation strategies.

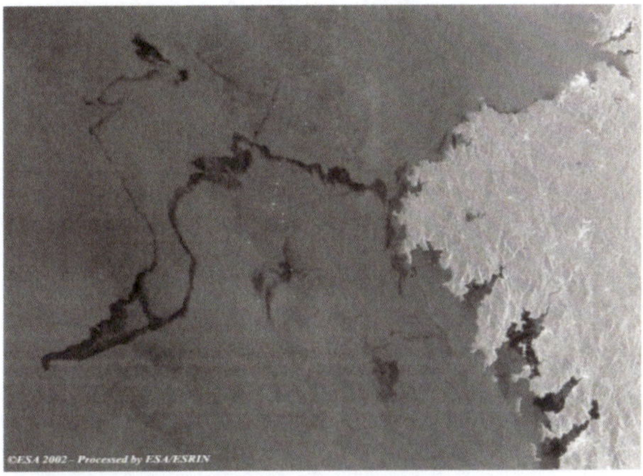

Figure 6. Left: Image showing the Prestige spills until its sinking (source: ESA).

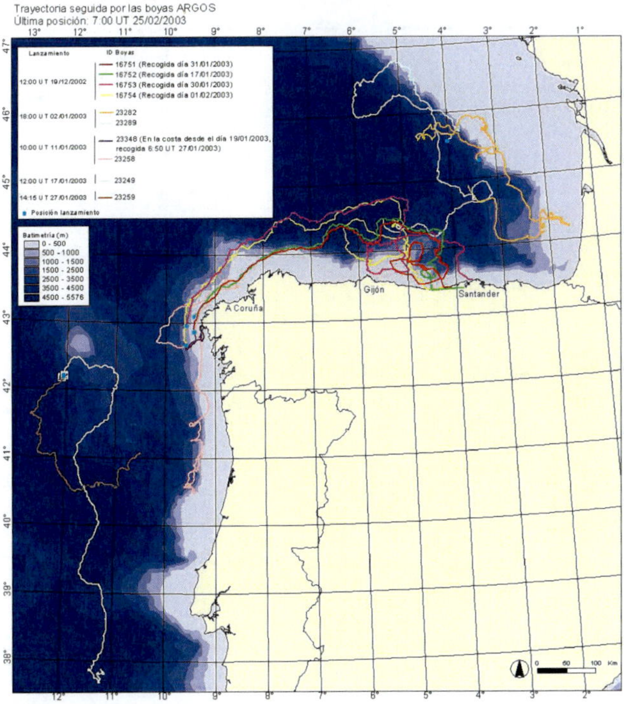

Figure 7. Overview of the three drifter sets during the Prestige incident (from García-Ladona et al., 2005).

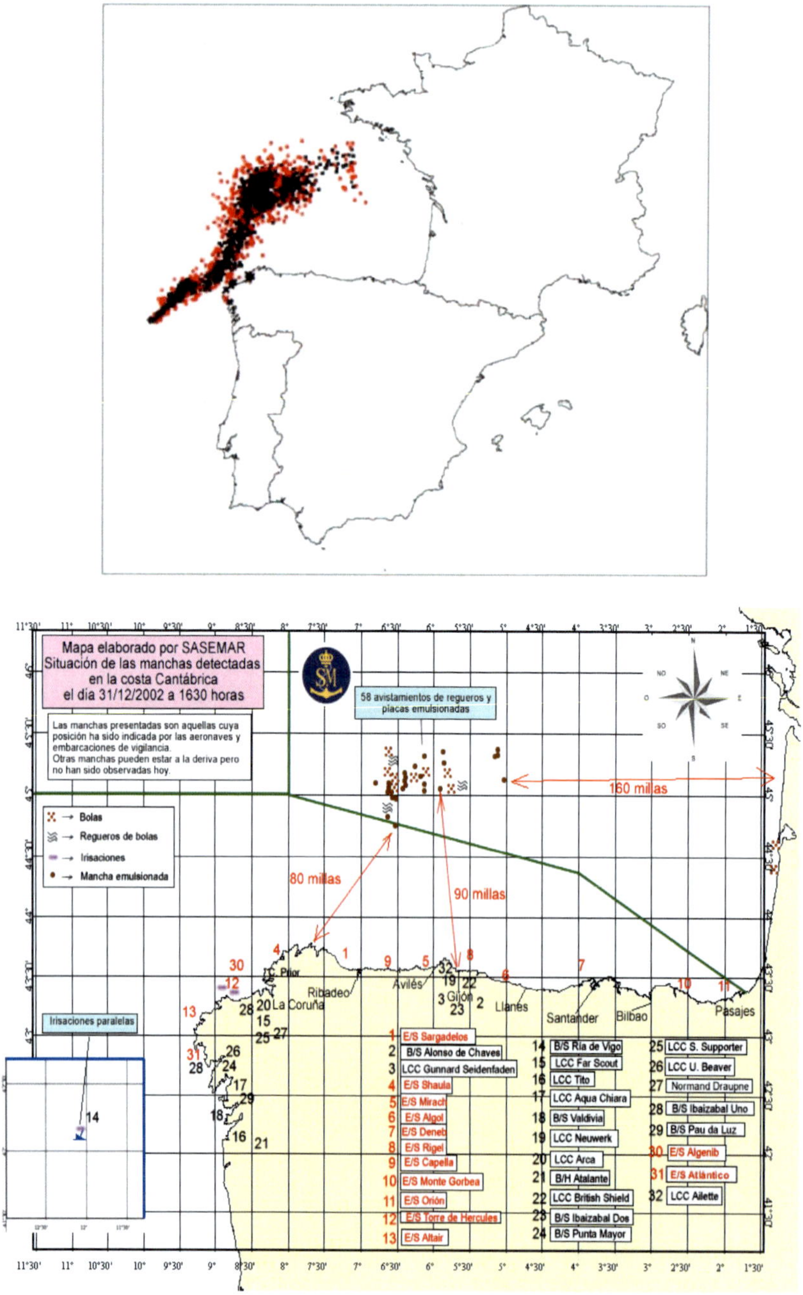

Figure 8. Top: GNOME long time forecasts. Bottom: Map collecting all the spills detected on 31/12/2002 (source SASEMAR).

Observational efforts to guide cleaning operations were concentrated in aircraft surveys to detect oil spills by visual inspection, to deploy Lagrangian drifters, during an oceanographic survey to establish the hydrographic conditions in the area and, to eventually track the biggest oil spills. A set of 10 drifters distributed in three groups were deployed. A first set was sent to track a group of oil spills located in front of Cape Finisterre (northwestern tip of Spain), a second was sent to track the area of the Bay of Biscay upon request from French authorities and a third group was sent to monitor the drift from the sinking point to four months after the shipwreck (Fig. 7). The first set moved following the shelf/slope topography entering the Bay of Biscay and after several loops arrived on the Cantabrian and Basque coasts (around 5-6ºW). The second group followed northward along the Aquitania shelf and reentered the deeper zone in the Bay of Biscay before ceasing to emit. The last group of drifters just moved southward because their release was done at the beginning of the upwelling season, when dominant winds favor an equator ward flow in the region. The ability of drifters to remain close to an oil spill has not been verified and is still lacking. However, an almost simultaneous arrival of spills and drifter parts were observed close to Santander near 6º W and was also in agreement with short-term forecasts (García-Ladona et al. 2005, see their Fig. 6). The singularity of this accident was an example of the difficulties in dealing with accurate oil spill dispersion forecasts (i.e. Daniel et al., 2003, Carracedo et al., 2006). A short-term prediction was needed when oil spills were detected close to the coast but a long-term prediction was also needed because a big spill originated far away from the coast when the ship sank. At that time, simulations predicted the main spreading of the spill toward the northern part of the Bay of Biscay, while in fact most of the spill flowed along the northern Spanish coast even reaching the Basquian coast in the southern corner of the Bay of Biscay. However, the area was so big that it cannot be disregarded that part of the spill, after its fragmentation, could have been advected according to such a forecast. In fact the map of Fig. 8 (right) shows the observations of the survey made on December 31st where oil bullets and patches of emulsified oil were detected at locations slightly to the east of the prediction made for the first days of January 2003 (Fig. 8, left).

Oceanographic reanalysis made after the Prestige accident allowed experts to analyze the influence of several aspects over the fuel dispersion during the accident (Carracedo et al., 2006). The path of the main spill was more in agreement with the aircraft survey for both the surface and the subsurface spills when climatological baroclinic currents were considered in the dispersion model. This is particularly important for the kind of heavy fuels that rapidly emulsify and may sink to some depth, so its dispersion can be much more directly influenced by the ocean current profile than the wind drift action. Another attempt to provide long-term forecasts during the Prestige incident was done by using the MOTHY system (Daniel, 1997, Daniel et al, 2004). Daniel and Dandin (2005) provided a reanalysis of the Prestige oil spill using two different ocean models: MERCATOR and FOAM. The lack of accurate velocity values in the first meters below the surface from such operational systems was solved by extrapolating the baroclinic currents at lower depths up to the surface and adding it to the wind drift effect. Although this procedure resulted in an improvement of the long-term behavior compared to GNOME and MOTHY alone, the problem lies in the differences that one can find between the model output fields.

2.2. The "Haven" Incident

The Haven incident is the most devastating environmental disaster that ever occurred in the Mediterranean (Fig. 9). On the 11th of April at about 12.30pm, the VLCC (Very Large Crude Carrier with a capacity of 250,000 tons of crude oil) "Haven" (Cypriot flag), was at anchor near the Porto Petroli of Genoa Multedo when it caught fire. The fire lasted for three days and finally the tanker exploded and broke into three parts. It was loaded with 144,000 tons of Iranian crude oil, which was released into the environment, and one of the tanker pieces sank on the spot. The two others sank during towing by tug ships. Despite the intense response from Italian authorities, oil slicks drifted westwards, hitting various parts of the coast of Liguria and then reaching the French coasts as far as Hyères. On the 14th of April, the slick was estimated to be about 12,000 m long and 4,000 m wide. But its drift was limited due to the lack of wind and calm sea conditions. Booms were rapidly deployed to protect the beaches from compact slicks. Recovery operations at sea and mechanical clean-up on the shoreline then began. Nearly 200 m^3 of pollutant were collected at sea. In total, almost 110 km of coastline, rocks and beaches were cleaned up. On the 15th of April, the French pollution contingency plan (Polmar plan) was activated. A study of the dynamics of the slicks showed a high rate of evaporation, but also the formation of bullets of oil. A surveillance network was set up to assess the quality of the air, seawater, beaches, coasts and marine flora and fauna. The immediate ecological consequences observed were the oiling of around 100 birds and the degradation of marine life in the surface layers and on the shore. In the long-term, it was essential to monitor the affected area so as to pick up on any potential deeper impacts on marine life which would only become apparent later. The impact on fauna was significant: a 43% reduction in fish populations was estimated in certain fishing areas.

Even today, small quantities of hydrocarbons occasionally leak from the principal wreck, and in a large seabed area, there are still several tar deposits. That necessitated a decontamination intervention on the principal wreck of the Haven; it has been financed through the compensation funds transferred by the Ministry of the Environment to the Liguria Region, for the purpose of eliminating the risk of leakage of hydrocarbons and oils (both fuel and lubricating oils), caused by the corrosion of the materials and the subsidence of the wreck's structures.

Oceanographic reanalysis (hindcast) of this incident was done by Daniel et al, (2003) using the MOTHY forecasting system. A two-week forecast using only the pure dispersion rates, according to very calm conditions over the area during the event, was unable to deal with the observations made by aircraft inspection. Several spillbullets were observed southeastward toward the Gulf of Genoa while the forecast only provided for a cloud around the accident locations. When the model just included an estimation of seasonal currents in the area derived from geostrophic approximation using a climatology atlas the results were greatly improved.

Figure 9. Left: The Haven tanker. Center: A view of the shipwreck (33-83 m depth). Right: the smoke and fire after the explosion.

2.3. The "Deepwater Horizon" Spill

A large spill occurred in the Gulf of Mexico when, on the 20th of April 2010 along the Louisiana coast of the US, the oil-platform Deepwater Horizon (MACONDO/MC 125) suffered an explosion followed by a fire. The rapid action by the US authorities rescued the survivors and deployed a large number of antipollution fighting apparatus. Two days later the oil-platform sunk releasing about 2,500 cubic meters of oil. Observations made with the help of ROVS revealed a continuous spill emanating from the rig well. From April to mid-July the well leaked a quantity estimated at about 4.9 million barrels (20 times the Exxon Valdez spill and 16 times the Prestige spill), which has been considered "the worst environmental disaster America has ever faced" (quoted from National Commission on the BP/Deepwater Horizon Oil Spill and Offshore Drilling, 2011, p:143) unprecedented in size, location and duration (Fig. 10).

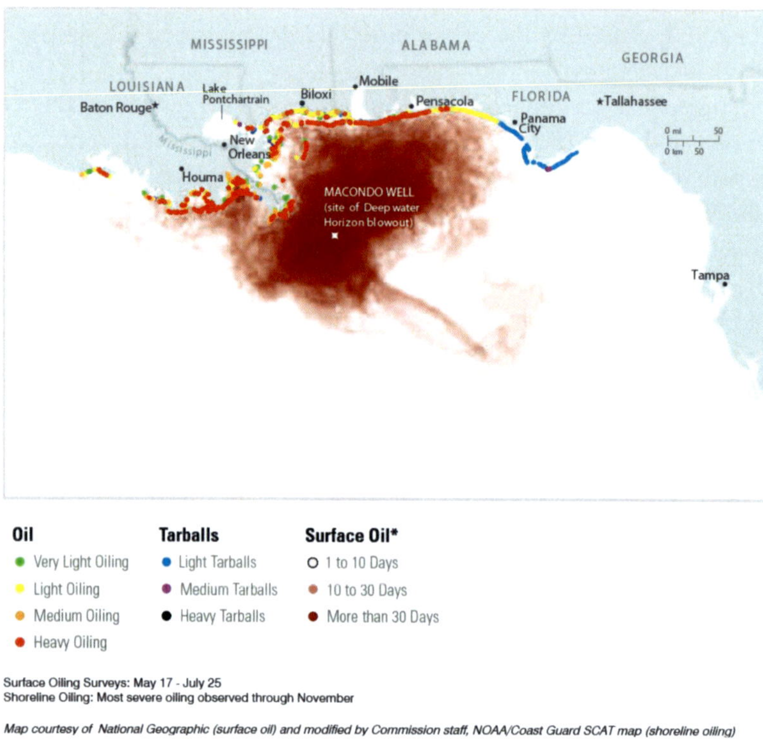

Figure 10. Maximum extent of oil in the Deep Water Horizon spill.

Given the characteristics of this spill it was challenging to perform an accurate medium and long-term forecast. First, being a deep source continuously releasing oil, the quantity spilt was even more uncertain with respect to other events. Second, besides the high pressure discharge and the positive buoyancy, some of the oil did not reach the surface around the well area but dispersed in deep layers as was observed and reported by Camilli et al., (2010). This means that part of the oil spill transport that occurred was undetectable by common surveys such as surface aircraft or remote sensing. In response, up to six ocean circulation models were available to make oil dispersion forecasts, which were combined into a sort of ensemble modeling. No single model was by itself adequate for all the different areas involved (the shelf, deep sea, estuarine regions…) but they provided early-warning information otherwise unavailable (Liu et al. 2011).

These three examples summarize the challenging aspects associated with devastating oil spills. The configuration in each case is quite different and so the actions to be taken are constrained accordingly. In the MT HAVEN spill, the environmental circumstances, good weather conditions, local proximity, and the partial oil consumption due to an accidental fire facilitated a fast reaction and a relatively low impact aside from the great amount of oil involved. The Prestige spill occurred in a relatively offshore location with a continuous release over an extended area due to the ship's path and subsequent collapse. Due to extremely bad weather conditions the operations were difficult to organize. As a consequence, the impact was enormous over a very wide area, though the quantity spilt was not as great compared to other spills. The Exxon Valdez was somehow a mixture of both events. It shared the same qualitatively environmental conditions as the Prestige spill, while being very close to the coast, as in the MT HAVEN spill, somehow limiting the relative extension of the impact. The Deepwater Horizon spill introduced a new element in respect to past events since the spill was located on the sea floor. From these and other past accidents, the need to improve current early-warning actions in terms of forecasting is necessary to enhance crisis management.

3. Review of Technical Approaches

3.1. Contingency Plans

Contingency plans are fundamental elements of most of the protocols required for organizing a quick response appropriate to the level of an emergency. They are often focused on the chain of activities and the administrative organization that should be set up in the event of an emergency. Although international agreements and regulations exist that provide a framework to homogenize the necessary contingency plans for maritime emergencies (see for example REMPEC information resources) the technical details and how they are implemented are dependent on the legislation of each country. Thus, the need to have contingency plans at the national, regional and local levels to be activated accordingly to the seriousness of an emergency is clearly established. Even supra-national regional plans are sometimes established between neighboring countries to activate common plans (i.e. RAMOGE plan between France, Monaco and Italy and LION PLAN between Spain and France). Nevertheless the *Search and Rescue (*SAR) services within the different countries have distinct organic responsibilities and proceed with different operational tools which can be a source of uncertainty in the decision making during a crisis. Often the technical aspects are not specifically detailed because they are dependent on the kind and complexity of the circumstances of an emergency. However, in search and rescue operations, pollution events and in many other emergencies, the basic questions are: when, where and how did it happen and what can be expected in the near future (Scholz et al., 1999). In many situations the environmental data are crucial to making decisions on the most appropriate response actions to be taken. The sea state (wave and currents) and weather (wind) conditions are the most general and obvious environmental conditions related to marine emergencies in oil spill events. For example, under strong wind conditions the wind drift would be important and should be modeled to prepare search and rescue operations of shipwrecked persons and objects. Big past pollution events, as the cases reviewed previously, constituted a challenge for predicting how the spills will evolve and arrive on the coast, and this task was necessary for short, medium and long-term periods. As a result, SAR agencies and authorities should have easy access to environmental data according to different scenarios, and eventually specific products adapted to their requirements.

There are two sources of environmental information: observed and simulated, although there is not a strict separation. In oceanography, the observational data rarely relies on a dense and well distributed network of instruments providing information regularly as it happens with the meteorological network of stations. In the ocean counterpart there are some national observational networks along coasts and harbors, open ocean arrays (i.e. TAO/TRITON, PIRATA), and some intensive observational programs (i.e. ARGO). Satellite data are perhaps the most regular way to cover wide regions synoptically with high resolution from space, but there are still sampling limitations. Ocean models are thus a useful alternative solution to provide ocean state data in a regular way. Besides the rapid development of operational oceanography in the past two decades (i.e., GOOS program) which has led to a spectacular development of the present ocean models as the best assessment tools in case of maritime emergencies, in general, an accurate sea state forecasting is still a long way off.

A forecasting system for oil spill assessment has to deal with two main parts: the observational and the modeling part. The observational part accounts for the necessary variables that are required to initialize the forecasts. The modeling part is usually composed of a weather forecast system, an ocean forecast system and an oil spill model. Oil spills models may be approached in two ways: by means of an Eulerian approach, considering the changes in time of the concentration of a substance at a fixed point or, by means of a Lagrangian approach considering the oil spill as an ensemble of particles (spillets) with a fixed mass or volume that are transported by currents.

Presently most of the commercial and non-commercial models rely on the Lagrangian point of view. In any case, a problem lies in the complexity of all the processes involved and the great number of parameters needed to represent them. Common approaches conceptually divide the processes into two main parts: transport processes and fate processes (weathering, e.g. see Sebastiao & Soares 1995; Scholz et al. 1999). Some mechanisms of weathering contribute to the disappearance of oil from the sea surface: evaporation, dissolution, vertical migration across the water column, adsorption and sedimentation, biodegradation; other weathering processes, such as emulsification, increase the oil persistence on surface; and other processes such as photo-oxidation have a dual role. Oils which are normally classified as persistent include crude oils, fuel oils, heavy diesel and lubricating oils. Non-persistent oils include gasoline, light diesel oil and kerosene. The persistent oil can be carried hundreds of miles in the form of scattered tarballs by winds and currents. The tarballs will vary in diameter from several meters to a few centimeters.

Spilled oil can float suspended in the water column or, as lighter hydrocarbons evaporate, the oil can become heavier, and sink. Small changes in water density may dictate whether the oil will sink or float, especially for dense oils. Dense oils may also sink after the incorporation of particles coming from bottom sediments, even though it may occur only very close to the coast. The wax and asphalt content influence the likelihood that the oil will mix with water to form a water-in-oil emulsion. Oils that form stable water-in-oil emulsions persist longer at the water surface, up to one year. The reader may consult the references of Fay (1971), Mackay & Patterson (1980), Delvigne & Sweeney (1988), Mackay& Leinonen (1977), Payne et al. (1984), and Hines & Maddox (1985), to understand the details of the empirical formulation of all such processes.

The mechanical transport part essentially includes three deterministic terms associated with the drift components due to wind, waves and ocean currents and a stochastic term representing dispersion due to the turbulent nature of the ocean flows (Diagram 1). We do not pretend here to make an exhaustive review of all the dispersion models and systems, but to illustrate the common characteristics. Thus, the examples below are representative of the kind of dispersion models and systems that are often adopted in several countries by official administrations and have been used in real cases.

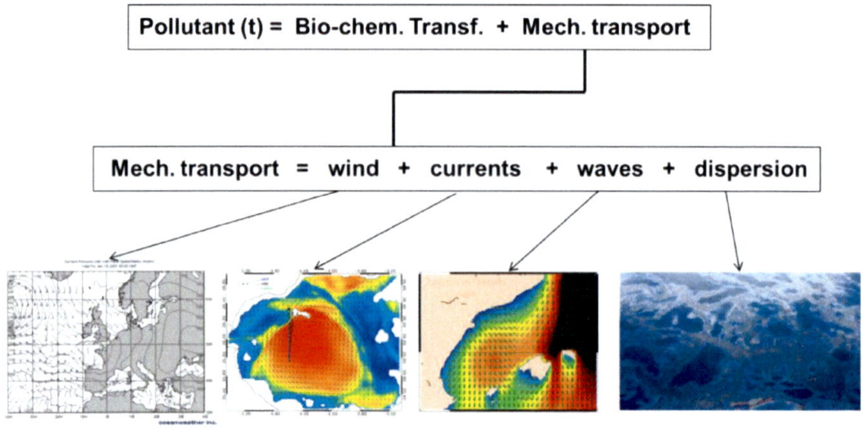

Diagram 1.

3.2. Lagrangian Approach to Mechanical Transport

The basic approach to mechanical transport from a Lagrangian point of view consists of considering spills as formed by a number of discrete particles (spillets) that are mechanically transported by a linear superposition of wind and wave drifts, the surface ocean current, and stochastic dispersion. If we consider a number of particles (n = 1,2,…,N) located at $\boldsymbol{x}^n = (x^n, y^n)$ at time t, their evolution may be computed by the following expressions:

$$\frac{d\mathbf{x}^n}{dt} = \mathbf{U}^n_{AD}(\mathbf{x},t) + \mathbf{U}^n_{ST}(\mathbf{x},t) \tag{1}$$

$$\mathbf{U}^n_{AD}(\mathbf{x},t) = \mathbf{U}^n_{CU}(\mathbf{x},t) + \alpha\mathbf{U}^n_{WA}(\beta\mathbf{x},t) + \gamma\mathbf{U}^n_{WI}(\mathbf{x},t) \tag{2}$$

where "*AD*", stands for advection and "*ST*" for stochastic dispersion. The advection is split in three terms according to the three main factors: wind (*WI*), waves (*WA*) and currents (*CU*). Without going into the mathematical representation of each term it is clear that to solve the transport part of a dispersion model we basically need the following variables: the wind velocity field, the wave field, the ocean surface velocity field and a way to parameterize the turbulent diffusion term. As our ability to pinpoint these variables improves, it will enhance our dispersion model or forecast system's ability to approach the processes in a more realistic way. However we have to contend with a stochastic part that usually represents the action of non-resolved motions at the resolution as we are integrating the system.

The strategies are diverse but we can separate them roughly in two groups: either we obtain the information from general meteo-oceanic forecast systems or directly from available observational data. In the first case, emphasis is put on forecasting models and in the second case, we should consider the best sampling strategy to get the necessary data. However, this separation is not strict because forecasting systems are to some extent based on existing data. Assimilation is the term used to indicate that a given forecast is through taking real-time or delayed observational data to improve it.

There is considerable literature on oil spill models. Often, the development of such models runs in parallel with the occurrence of large oil spills, as was the case with the Prestige accident. Individual and uncoordinated initiatives by several research groups were set up to provide some guidance. In particular there were attempts to forecast the spills by research groups of different institutions and countries separately (e.g., Balseiro et al., 2003; Carracedo et al., 2006; González et al., 2006; Daniel et al. 2004; Daniel and Dandin, 2005; Daniel et al. 2005) that have evolved into more sophisticated forecasting systems.

The MOHID system (http://www.mohid.com) is one example. It is not strictly an oil spill model but a three-dimensional water modeling system, developed by MARETEC (Marine and Environmental Technology Research Center) in the Instituto Superior Técnico (IST) which belongs to the Technical University of Lisbon (UTL, see references in http://www.mohid.com /Publications/JP.asp). The MOHID system allows the adoption of an integrated modeling philosophy, not only for the processes (physical and biogeochemical) and scales (allowing the use of nested models) but also for systems (open sea, estuaries, watersheds) due to the adoption of an object oriented programming architecture (Miranda et al, 2000). For oil spill modeling, it contains a Lagrangian module linked to an oil-weathering module to incorporate physical-chemical transformation processes (Fernandes, 2001). An updated implementation of the system successfully reproduced the main features of the first Prestige oil spill (Carracedo et al., 2006).

The MOTHY (Modèle Océanique de Transport d'Hydrocarbures) model developed by Météo-France is used on an operational basis to predict surface/subsurface transport of spilled oil, container drift and search and rescue targets. It is based on a limited domain ocean model coupled to a pollutant dispersion model forced by wind and pressure fields from the French weather forecasting system. The oil slick is modeled as a distribution of independent droplets that move in response to currents, turbulence and buoyancy. Turbulent diffusion is modeled with a three-dimensional random walk technique. The buoyancy force depends on the density and size of the oil droplets so that larger (more buoyant) droplets tend to remain in the surface layer whereas the smaller droplets are mixed downwards (Elliot, 1986). If a droplet is moved on to land, then that droplet is considered beached and takes no further part in the simulation.

TESEO was the model developed during the ESEOO Spanish project for oil spill forecasting (Alvarez-Fanjul et al. 2007). The model was a new implementation of an original PICHI model developed by the University of Cantabria during the Prestige incident (Castanedo et al., 2006, Abascal et al., 2008). The model allows the total integration of atmospheric and oceanographic forecast fields (standard netcdf formats) and optionally can be run with a manual input of characteristic or observed wind and sea state fields maintained unmodified along the integration. A particular characteristic compared with most of the oil spill models is that drift/drag coefficients of the different mechanical transport components are determined by maximization of a cost function that weights forecast and external data (usually drifter trajectories). Thus coefficients are not fixed at all during the time integration but are time and space empirical varying functions updated at some intervals. The model was used during the ESEOO exercises (Sotillo et al. 2010) and a calibration working with different maximization schemes has been published using the drifter data collected during the Prestige accident (Castanedo et al. 2006; Abascal et al. 2009 b; 2010).

The GNOME (General NOAA Oil Modeling Environment) (Beegle-Krause, 2001; GNOME 2002; Beegle-Krause et al. 2003; Beegle-Krause 2008) is a free code that can be used to forecast the spread of oil spilled on water and its physic-chemical changes, and to learn how predicted oil trajectories are affected by inexactness ("uncertainty") in current and wind observations and forecasts. GNOME uses standard approaches to forecast the evolution of an oil spill's Lagrangian Elements (LEs or spill dots called *splots*). The model combines surface physics effects caused by winds - such as wave stress, wave compression, Stokes drift, dispersion, over-washing, surface drift, and Langmuir circulation - into a single drag parameter called the Wind Drift Factor. Default values for the Wind Drift Factor range from 1-4% of the wind speed although any values from 0-100% may be used. This allows the user to simulate the subsurface drift of submerged tarballs and other pollutants that would be unaffected by the wind, as well as objects that have higher wind drag than oil, such as drifting ships, large floating debris, and jetsam.

The OILMAP system (Applied Science Associates) is a commercial oil spill forecast system developed by a science and technology solutions company (Applied Science Associates, ASA http://www.asascience. com) which has been used and adopted by some national SAR services (e.g. Howlett et al., 2008 and references within http://www.asascience.com/about /publications/index.shtml). The OILMAP is a characteristic oil spill model composed of a trajectory model and a weathering model that is able to predict the subsurface transport of entrained and dissolved oil. The model has been tested, calibrated and used in several oil spill and pollution incidents (Spaulding et al. 1992; Spaulding et al. 1994; French-McCay et al. 2008; French-McCay 2009). Presently the model is integrated in more general tools such as SIMAP, which combines an oil spill model together with an ecological model to allow a complete impact assessment of an oil spill (French-McCay and Rowe 2004).

MEDSLIK is an oil-spill model developed at the Oceanography Centre belonging to the University of Cyprus in Nicosia (Lardner and Zodiatis, 1998; Lardner et al., 2006). MEDSLIK also consists of a drifting model with a representation of transformation processes to simulate the fate and dispersal of oil slicks. It has been recently used to carry out a hindcast study of the Lebanon oil spill in July–August 2006 (Fig. 16, from Coppini et al. 2011). MEDSLIK has evolved into MEDSLIK-II tool which is now an open source version of the original code that deals with a consistent representation and coupling of weathering processes and mechanical transport affecting oil spills particles (De Dominicis et al. 2013 a, b). This version has been selected to be the core of a multi-model oil spill system developed on the frame of the MEDESS-4MS project (INTEREG MED 2S-MED11-01) to be operational for the whole Mediterranean basin.

Finally, during the TOSCA project (INTEREG, G-MED09-425) a Lagrangian model for tracking surface spills and SAR operations in the ocean has been developed (Sayol et al. 2014). The main characteristic compared with other Lagrangian models is that it allows spatial and temporal diffusivities to be included in the stochastic dispersion part to account for the inaccuracies of the ocean fields. These terms are computed from a statistical analysis over a certain region of interest of the past ocean velocities forecast (typically 3 years). The final result is given in terms of the probability density function of an ensemble of particles released at the initial location.

There are many other models available with essentially similar characteristics to those briefly quoted before (e.g. GULFSPILL, Al-Rabeh et al. 2000; SINTEF-OSCAR, Reed et al. 1995). It is worthwhile to mention the model presented by Wang and Chen (2010) that incorporates three-dimensional transport terms in unstructured meshes with diffusivities coming from the ocean circulation models through Smagorinsky's formula. The vertical diffusion is computed using a Langevin model of the first order. Comparisons with some analytical and synthetic experiments are straightforward but have not been tested in real cases.

A common aspect of the Lagrangian approach is that the forecast abilities are highly dependent on the accuracy of the Eulerian ocean currents. Thus, Coppini et al. (2011) found that the same oil spill code, coupled to two Eulerian velocity fields of different resolutions, performed distinctly in the case of the Lebanon spill. The results were in better agreement with observations when the higher resolution model was used, in this case because it improved the assessment of the coastal area where the spill occurred. Twin experiments with the CAMCAT oil spill forecasting system showed that errors in the wind and current fields are the main limiting factor for the quality of the forecasts (Jordà et al., 2007).

To determine all the terms of equation 2, the components which essentially come from a (regular or irregular) grid, have to be interpolated in space and time at the particle location. To do this, many models use the bilinear interpolation, which is the simplest interpolation method on a surface.

Bilinear interpolation is often accurate enough to study the general behavior of the field. As the interpolating point moves from a grid cell to its neighbor, the interpolated function varies continuously. However, the gradient of the interpolated function changes discontinuously at the boundaries of each grid square. These discontinuities could produce artifacts and weird behavior in the results provided by the models.

Thus, it is recommended to use higher-order methods to obtain increased accuracy for the interpolated function and improve the continuity of the gradient. In this regard, "Bicubic interpolation" is suggested to provide a value for the field outside of the grid points. Bicubic interpolation requires specifying the function value, gradients and the cross derivative at each grid point. The final interpolating function satisfies the following properties: (i) The values of the function and the specified derivatives are reproduced exactly at the grid points, and (ii) the values of the function and the specified derivatives change continuously as the interpolating point crosses from one grid square to another. The equations of bicubic interpolation do not require you to specify the extra derivatives correctly as the smoothness properties are forced by construction of the method. However, as you provide better required values, the interpolation will become more accurate (the interpolating function will be smooth in any case). It is easy to compute all the derivative values required by this interpolating scheme by using centered differencing.

Also, to interpolate the field in time, for the same reasons, using a better method than the linear interpolation is recommended. There are several choices but finally the most appropriate is to use bicubic spatial interpolation in space and third order Lagrange polynomials in time. There are interpolating methods in time with higher accuracy but its computational cost is huge (as can be seen in the comparison made by Mancho et al. 2006); in contrast, the interpolation proposed gives accurate results at a moderate computational cost.

A second aspect often not well documented in most of the oil spill models is the time integration of the Lagrangian equations. For advecting particles, classical Runge-Kutta methods are recommended. The order "n" for the Runge-Kutta method means that the error per step in on the order of h^{n+1}, and the total accumulated error has order $O(h^n)$, where h is the step-size. Most of the models implement up to the fourth/fifth-order method using standard algorithms (Press et al. 1996). However, classical Runge-Kutta methods are convenient only when the dispersion/stochastic term are not present. When a stochastic term is introduced in the model, a stochastic Runge-Kutta algorithm should be constructed according to the statistical properties of the dispersion term in order to preserve them along the timeline.

The predictability horizon of forecasting models is very sensitive to the resolution of the data for the initial velocity field. With typical resolutions of atmospheric observation systems, meteorological models can predict up to 7 days ahead. Beyond 7 days, the prediction has a lower reliability than a random prediction. Given that ocean surface currents are partially moved by winds, the predictability horizon of the ocean velocity field is not much longer than 7 days.

Given that wave drag frequently correlates with the local wind, the oil slick movement on the ocean surface is frequently modeled by the following expression:

$$v=\left(ux+\frac{\tau x}{f},uy+\frac{\tau y}{f}\right)\cong\left(ux+0{,}03(\cos\beta Ux+\sin\beta Uy),uy+0{,}03(-\sin\beta Ux+\cos\beta Uy)\right)$$

where f is the water-oil friction coefficient (Tkalich and Xaiaobo 2001; Zadeh and Hejazi 2012) and $\tau_{x,y}$ are the two components of the wind stress (N/m^2). Normally, a windage factor of 3% of the wind speed at 10 m over sea level is used, and the angle β (between 0 and 25°) takes into account the fact that the wind drag generates surface currents to a certain angle due to the Ekman (Coriolis) effects. Tidal effects are sometimes included as an additional term, but they only affect the oil drift near the shoreline. However, it may be important to model wave drag separately (Daniel et al. 2003) because the Stokes drift is not always aligned with the wind direction. This requires using realistic forecasts of wave fields which are not always available.

The experience obtained from international projects such as TOSCA indicates that, to obtain good forecasts from given initial conditions, the quality and resolution of the initial, meteorological and hydrological data are more important than the quality of the existing forecasting models.

Geophysical fluids (atmosphere and ocean) are turbulent fluids with nested structures of eddies at different scales. The usual resolution of hydrological velocity fields in the ocean is 0.125 x 0.125°. In the sub-scales under this resolved scale, many eddy structures remain and, when they are energetic, are able to change the local direction of the slick movement with respect to the direction predicted by the solved hydrological velocity field. To see the influence of the resolution and sub-scale structures, we can observe Figure 11, where up to four oceanographic models predict very distinct behaviors for particles released at three nearby points and are also compared with the real trajectories of surface drifters released over the same period.

Given that the physics equations used by such models are almost identical, the difference can only come from the different resolution and quality of the wind and ocean fields assimilated by the models.

To see the influence of the turbulent movements in the sub-scales, it is useful to compare the predictions obtained when these fluctuations are incorporated (or not) to the model. With this regard, we start assuming a given turbulent horizontal diffusivity given by the Okubo (1971) diffusivity (m^2/s) as a function of the scale L_r (m), which is given by the following expression:

$$K_x = 2.06 \times 10^{-4}\, L_r^{1.15} \tag{4}$$

L_T is chosen as 8640 m, which is a typical length of sub-scales that are not solved by the resolution of many hydrologic datasets.

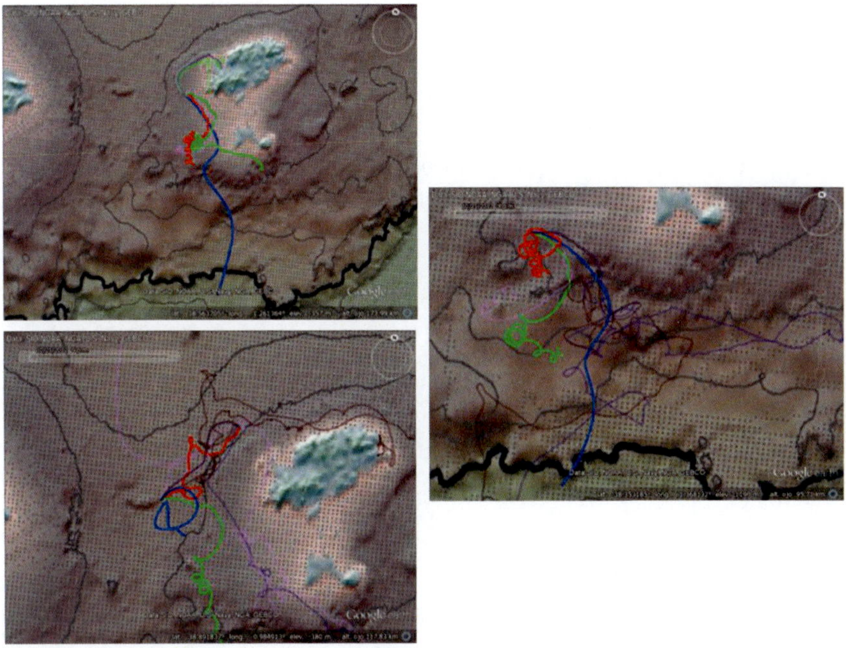

Figure 11. Pure advection of particles at the surface obtained by integrating the velocity fields from 4 different operational models of different resolutions (pink: ESEOO model, red: IBI-MYOCEAN model, blue: MED-MYOCEAN model, green: SOCIB-IMEDEA). Light lines correspond to real observations of surface drifters trajectories during the BALEARES experiment carried out during the TOSCA project.

The variance of the probability density function (PDF) of the surface fluctuating velocities, or "Turbulent Kinetic Energy" is assumed to be $V_x = V_y = 2.5\text{x}10^{-5}\ m^2/s^2$.

An Ornstein - Uhlenbeck process in two dimensions is implemented for the fluctuating velocities:

$$u_t = u_{t-1} + C_d\, u_{t-1}\mathrm{dt} + \sigma\, \mathrm{R}[\mathrm{N}(0, \mathrm{dt}^{0.5})] \quad (5)$$

$$v_t = v_{t-1} + C_d\, v_{t-1}\, \mathrm{dt} + \sigma\, \mathrm{R}[\mathrm{N}(0, \mathrm{dt}^{0.5})] \quad (6)$$

where *t* is a time index, *dt* is the size of the time step, R[*d*] is a random number of the distribution *d*, N(0,*s*) is the normal distribution of mean *0* and standard deviation *s*, and $c_d = -2V_x / K_x$ and $\sigma = 2V_x / K_x^{0.5}$ must have that functional form to make the PDF of the fluctuations steady. These fluctuating velocities are added to the corresponding advective velocities obtained by interpolation of the hydrological ocean data.

Figure 12 shows the mean distance between a particle which drifts from a given initial position and is subjected to turbulence and the same particle drifting without the effect of turbulence, as a function of time (blue line). A set of 441 particles, homogeneously distributed in a rectangle of 0.5° x 0.5° around the position (6°W, 41.5°N), was used to make the calculation and obtain the average distance as well as the standard deviation of those distances as a function of time (red line).

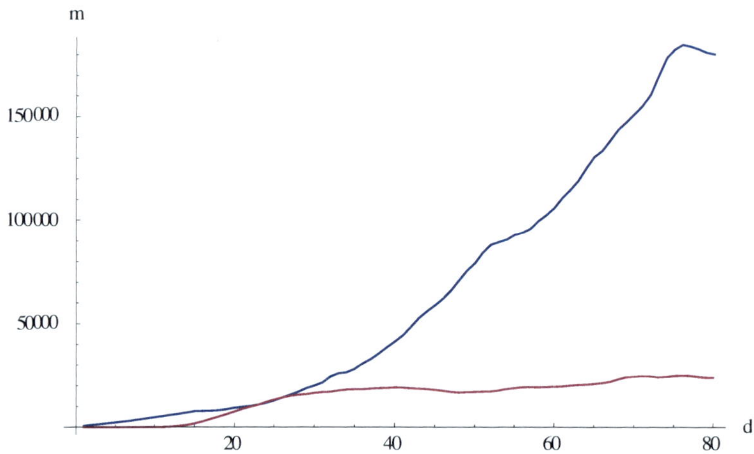

Figure 12. Mean distance (upper line) and standard deviation (lower line) of couples of particles drifting with and without turbulence across the Mediterranean Sea.

As can be observed, the effect of the sub-scale turbulence is to produce a rising separation between the trajectory with turbulence and the trajectory without turbulence. The mean separation rises exponentially during the first 50 days until a value of about 75 km. After that date, the trend is no longer exponential due to the rising number of tracers that become beached on the coast.

3.3. Observational Strategies

To obtain each of the components of the equation (2), operational oceanography forecasts should be the first choice if they are available because they are state-of-the-art tools for assimilating meteorological data, wave data and hydrography, they use state-of-the-art physics and give the best available representation of the turbulent velocity field. Operational systems include:

- nowcasts providing an accurate description of the present state of the sea, even at a higher resolution than the initial data sets,
- forecasts providing continuous estimations of the future condition of the sea for as far ahead as allowed by the resolution of the initial data sets,
- hindcasts assembling long-term data sets which will provide data for descriptions of past states, and their trends and changes.

In Europe, the European Global Ocean Observing System is one of the most important systems available currently oriented to these purposes (www.eurogoos.org). The equivalent system in the US is the Center for Operational Oceanographic Products and Services (CO-OPS) of the National Oceanographic and Atmospheric Administration (NOAA) (http://tidesand currents.noaa.gov/).

The MEDESS-4MS integrated system is being developed on the frame of a European Union project (MEDESS-4MS) which is currently setting up an integrated real time operational oil spill forecasting service for the Mediterranean area. In addition, it will provide support to the response agencies for improving their contingency plans and for implementing the 2005/35/EC Directive on identifying ship-source pollution. The service is specifically designed to serve the needs of the Member and non-Member States response agencies, as defined by the European Maritime Safety Agency (EMSA) and the Regional Marine Pollution Emergency Response Centre (REMPEC).

The objectives of this system are:

- To implement an integrated real time multi-model oil spill forecasting system based on five well established drift and weathering models using state-of-the-art technology (PORTICO-MEDSLIK, MOTHY, POSEIDON-OSM, MEDSLIK and MEDSLIK II).
- To implement an interconnected network of data repositories that will archive and provide in an operational way, access to all available environmental and oil spill data.
- To test the service functionalities with key end-users: REMPEC, EMSA, and national agencies responsible for combating oil spills.

- To develop the integrated system with a unique access web portal (http://www.medess4ms.eu/) with different services and user profiles, multi-model data access and interactive capabilities.

The multi-model oil spill forecasting system is composed of environmental information from the Global Monitoring for Environment and Security (GMES) Marine Service and the national ocean forecasting systems interfaced with oil slick data from existing monitoring platforms. It will use the real time information about position of the oil slick; interface it with oil spill models capable of forecasting the movement of the pollution providing tailored products to oil spill crisis management users. The service will be accessible from a single portal expected to be useful for operational oil spill predictions in the Mediterranean at regional, sub-regional and coastal scales.

However, aside from the rapid advance in oil spill modeling we are far from being in an optimal position. Indeed, most of the operational systems rely on data assimilation procedures. Without data from measurements the forecast quickly diverges. Today weather forecasting systems are very successful thanks to a dense network of meteorological stations and routine vertical profiling of the atmosphere's upper layers. The oceanographic counterpart is not equivalent. The ocean observation is limited by the high cost of maintaining an observational network. Remote sensing offers perhaps the most straightforward way to retrieve synoptic information, i.e., simultaneous information over a wide area. Although parts of these data are assimilated into ocean models, we cannot avoid the fact that the better the initial conditions the better the forecast and accuracy of any non-linear predictive model. Therefore it is advisable to make direct use of the field information provided by measurements as a primary input of the different terms in equation (2).

As we commented above, the meteorological information provided by the weather forecasting system is quite accurate for short-terms. The wave field can also be predicted rather well from wave modeling from the meteorological forecast, but the accurate estimation of the ocean velocity field is still a challenge. In this regard, different field measurements can be of great help to improve the quality of the initial conditions necessary to set up the oil spill forecast. In the following, we will identify some strategies that may be of great help in providing a first guess about the initial conditions of the velocity fields in a marine emergency. These have to be considered as a complementary use of ocean models, although in some cases, it may be the best available information at the time of the accident.

Synoptic measurements of ocean velocity fields covering a certain region continue to be a major technological challenge. Since the beginning, oceanographers have tried to develop instruments to measure it. The observational estimation of the velocity fields can be roughly divided into two categories: direct and indirect measurement. "Direct measurement" means here, the observation of a magnitude directly or univocally related with the velocity field. By contrast, "indirect measurement" means the obtaining of the velocity field from a related magnitude under some dynamical hypothesis or assumptions. We briefly illustrate these two methods.

3.3.1. Direct Measure of Velocity Fields

Practically all the modern measures of ocean currents are based on devices that exploit the acoustic or electromagnetic Doppler effect of signals scattered by suspended particles in the case of acoustic devices and the moving surface roughness in the case of High Frequency (HF) radars.

Acoustic devices are normally point-based systems providing the two or three velocity components (depending of the number of transductors) either at a fixed point (ADC) or over a range of depths of the water column (ADCP). In the first case, the information is limited in space because they are usually implemented through moored lines for monitoring over long time periods at a fixed location. More space intensive measurements can be undertaken with ADCP profilers that provide the velocity over the entire water column again at a fixed location. If this instrument is mounted in the hull of a vessel we can obtain a section of vertical profiles during its trajectory. This equipment commonly mounted in oceanographic research-vessels can also be installed in underwater autonomous vehicles as AUV or gliders. By joining several data sections a three dimensional view of the velocity field over a certain domain can be reconstructed. The main drawback is the need of a good positioning system to filter the individual motions of the carrier engine and the corrections when interpolating the data to avoid spurious contamination and aliasing of low frequency motions as tidal and inertial motions (i.e. Font et al. 1998; Ruiz et al. 2002; Garcia-Górriz et al. 2003).

Coastal radar systems are also used to retrieve the velocity field from the roughness of the ocean surface. Although there are two systems, CODAR (http://www.codar.com/) and WERA (http://www.helzel.com/de/6035-WERA-Remote-Ocean-Sensing) they are based on practically the same principle. The Doppler shift of electromagnetic waves reflected by a moving ocean surface allows for determining the velocity of the water parcels in the ocean surface. By combining the signal of two or more antennas, depending on the system, a two dimensional velocity field can be obtained over an area depending on the antenna range. In the CODAR system, a single antenna is enough although combining two or three are usual configurations, while the VERA system works with a line of antennas. The output is equivalent to a two dimensional sea surface velocity field over a given domain (Fig. 13).

These systems are often located in coastal areas and have typical ranges between 50 and 100 km. These systems are optimal for carrying a time dependent two dimensional monitoring of surface currents and may constitute important tools for monitoring oil spills in coastal zones (Abascal et al., 2009) During the TOSCA project intensive experiments combining drifter deployment and radar observations showed improvements in the forecasting of surface ocean currents. A prototype of a forecasting tool (TOSCA tool, Fig 12), that simultaneously combines observations from radar, drifters and models, is available for several regions of the Mediterranean basin (http://toscagis.univ-tln.fr/TOSCA/squel.php).

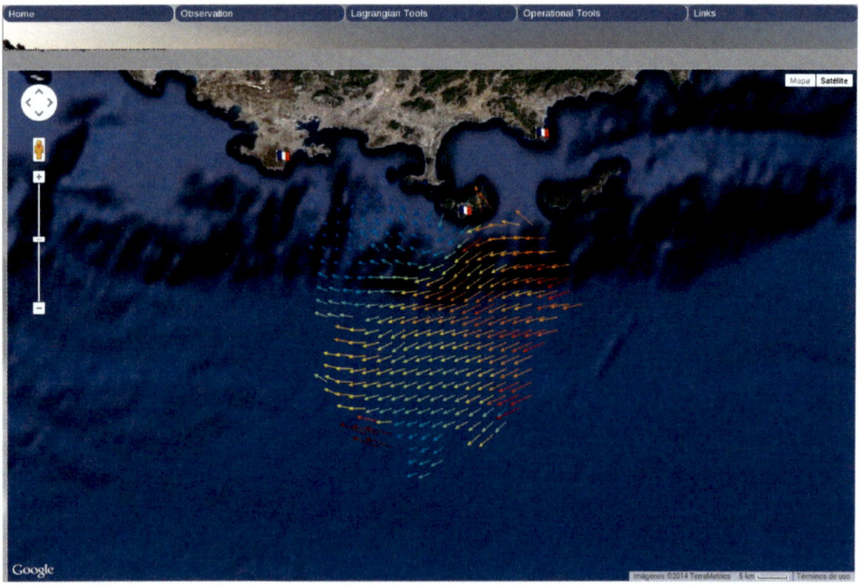

Figure 13. A look at the TOSCA tool developed on the frame of TOSCA project. The tool allows for simultaneous combining of real time trajectories from drifters, radar observations, model outputs including the output of an oil spill model if needed (Sayol et al., 2014).

Finally, the same principle can be applied to airborne transported devices and satellite remote sensors, which open the way to cover wider open ocean regions. Some attempts and demonstrations have been done using ASAR from ENVISAT images to retrieve and produce a field velocity from radar scenes (Johannessen et al. 2008).

Remotely tracking drifters is another way to get the water velocity almost directly. As illustrated in the Prestige case, their use for tracking oil spills are not new and have been analyzed since the beginning of 90's (Reed et al. 1990, 1994). Properly speaking they do not measure the water velocity directly but if we assume that they behave as a water parcel then their motion, and its associated velocity, may roughly approximate the velocity of water parcels. Drifters can be different shapes and configured in different ways (i.e. dragged or not), and may also be equipped with additional sensors (temperature, Doppler profilers, salinity, etc.). Their paths allow the water velocity to be computed by time differencing its position. The differences with previous methods are basically that the water velocity is computed along the track following the drifter motion while the previous systems basically give the velocity at a fixed point. Even if the information is of the Lagrangian type, Eulerian velocity fields – which means a map of approximately synoptic velocities— can be derived from drifter observations if we have enough space and time coverage (Poulain & Niiler 1989; Eremev et al. 1992). The statistical processing of trajectories under the constraint of physical assumptions has been the normal procedure for obtaining the dispersion characteristics of ocean flows (e.g. see De Dominicis et al. 2012 and references therein).

3.3.2. Indirect Measures of Velocity Fields

Geostrophic Approach

Before acoustic current meters were developed, ocean velocities were measured directly through mechanical devices often deployed in mooring lines. However the most classical approaches to obtaining ocean velocities were mostly based through indirect methods derived from assumptions about the ocean dynamics. These strategies have been at the core of ocean research since the beginning. The fact that ocean dynamics, as a first approximation, can be explained as being in geostrophic equilibrium, namely balance between pressure gradients and Coriolis acceleration, has allowed scientists to derive velocity fields from geostrophic relationships. Without going into details let's just say that if we are able to determine the density and the surface shape of the ocean water column we can estimate the ocean velocity field. We basically need to measure the temperature and salinity (in fact conductivity) and then use the seawater state equation to compute the density. Such variables can be measured with the aid of typical CTD (from Conductivity, Temperature and Depth) profilers lowered from the surface to high depth. CTDs can also be incorporated into autonomous vehicles (AUV and gliders) and profilers, such as the ARGO buoys program (http://www.argo.ucsd.edu/Novel_Argo.html). The reason for using all these devices is the same, to retrieve vertical profiles of such variables over an area, compute the corresponding densities, interpolate them spatially and finally derive the velocity field using the geostrophic approximation. Autonomous vehicles and ARGO-like profilers can do that continuously along their pathways. Data can be regularly transmitted by satellite if the vehicle periodically surfaces.

Even from a limited amount of data it is also possible to have an estimation of the density structure in a given area by combining previous archived data and expendable profilers (i.e. XBT, XCTD, http://www.sippican.com/). These are devices than can be easily deployed from a ship without any special infrastructure as it occurs with a CTD. Although CTD equivalent expendable profilers exist (XCTD) they are quite expensive. Less expensive expendable profilers, which only provide the temperature profile (XBT), can be used and combined with historical data to retrieve an approximation of the vertical structure of density. If we dispose of climatological data from a region sampled with XBT, empirical T-S relationships can be used to get the associated salinity profile and then, to derive the density and geostrophic field. The method can be as simple as having polynomial fits to historical T-S diagrams or more elaborated and sophisticated fits using vertical Empirical Ortogonal Function decomposition of the historical data or even using trained neural networks. The literature on such methods is quite extensive (e.g., Emery & Dewar 1982; Thacker and Sindlinger 2006, and references inside). An advantage of the XBT compared with the CTD cast sampling is, that they can be operated from virtually any kind of vessel, do not require stopping the vessel and thus, can be easily embedded within oil spill emergency operations. The disadvantage is that the representativeness of the reconstructed density profile is limited and may not reflect in detail the real conditions. For example, with such purposes in mind, Machín et al. (2008), carried out a procedure for an XBT sampling done using a fast search and rescue boat in an simulated emergency in the Gulf of Biscay (Northern Spanish coast). From a set of XBT profiles they fitted climatological T-S profiles from historical data in the sampling area and used it to infer the corresponding velocity profiles (Fig. 14).

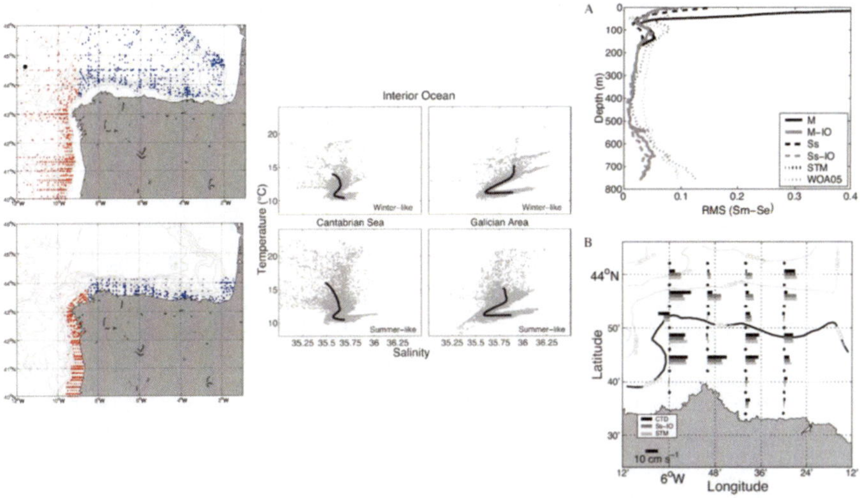

Figure 14. Right: Historical CTD data in the Northwestern Spanish coast. The data set was split into deep and shelf data sets (upper and lower map respectively) and into west and north regions. Center: polynomial T-S fits to historical data taking into account seasonality. Left: vertical profiles of RMS geostrophic velocities for different strategies (upper, A) and derived surface velocity vectors in the sampled region (lower, B).

Another source of velocity fields derived indirectly is from radar altimetry from satellites. The altimetry can provide the shape of the ocean surface, then the gradients of the surface shape are related with pressure forces over the water column and the geostrophic relationship can again be used to derive ocean velocities. Present altimeters only provide the surface shape along their tracks so only the gradients that are aligned with these tracks are directly observable. With the combination of various satellites and different orbital tracks, a two dimensional ocean velocity field can be derived. However, the spatial-temporal extents are limited by the information retrieval and may not be changed during an emergency. Despite the set-up, altimeters have revolutionized oceanographic research, the resolving scales of the ocean mesoscale can only be adequately resolved when several of these instruments are combined which is not often the case (Pascual et al. 2006). The information is good enough for undertaking ocean research, but far from being suitable for direct operational purposes. Usually such information is used for data assimilation purposes within present operational models.

Features tracking velocities: By feature tracking velocities we mean velocities estimated by recording the motion of a feature that can be tracked in space and time. A feature can be something similar to a patch of surface warm/cold water that is advected by the surface currents (eddies, filaments, etc.) or an iceberg drifting with the ocean current. In the atmosphere the motion of clouds in a sequence of satellite images can be used to infer the wind velocity (e.g., Davies et al. 2007). Similarly in the ocean, a patch of water of contrasting properties observed in a sequence of infrared or visible images may allow the retrieving of a velocity field. The most common technique is referenced as Maximum Cross-Correlation technique (MCC) developed in the 1980's (Emery et al., 1986; Garcia and Robinson, 1989). It is based on finding similar pattern structures through consecutive images according to a maximum correlation of its shape. Operationally this technique has many limitations related to the availability of a sequence of images, cloud contamination and other drawbacks related with the data processing algorithms. Nevertheless this technique can still be useful when complemented with other sources of information which overtake such difficulties; for example, combining it with other surface satellite data (Fig. 15, Crocker et al., 2007) or Lagrangian observations and radar systems (Notarstefano et al., 2008).

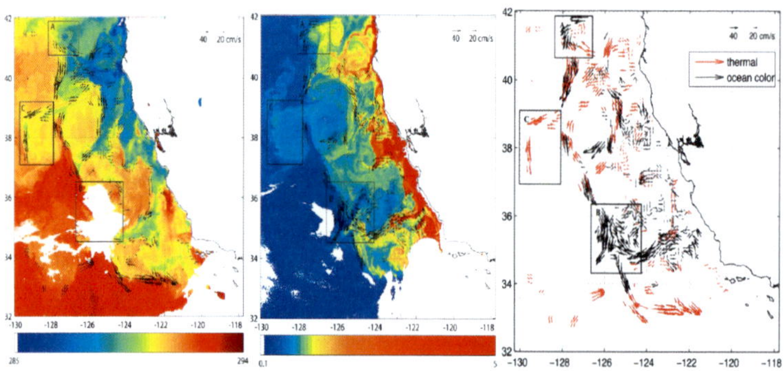

Figure 15. Left: thermal derived MCC vectors from infrared images. Center: vectors derived from Chl-a image. Right: composite of both velocity fields (from Crocker et al., 2007).

There are new approaches based on exploiting the geometrical shapes of surface patterns seen from satellite images that also may allow inferring the surface velocity field. Recently the Maximum Singularity Exponent technique (Turiel et al., 2005, 2008) may have been used, as a proxy for the ocean stream function but additional external data is required to provide consistent velocity fields. This technique relies on the properties of the Maximum Singular Exponents that give an idea of how tracer gradient scales with the resolution in a fully developed turbulent flow. The exponents of the scaling characterize the skeleton of the most dynamically relevant flow patterns. The lines associated with these singular exponents are closely related to the geometry of the velocity fields (Fig. 16).

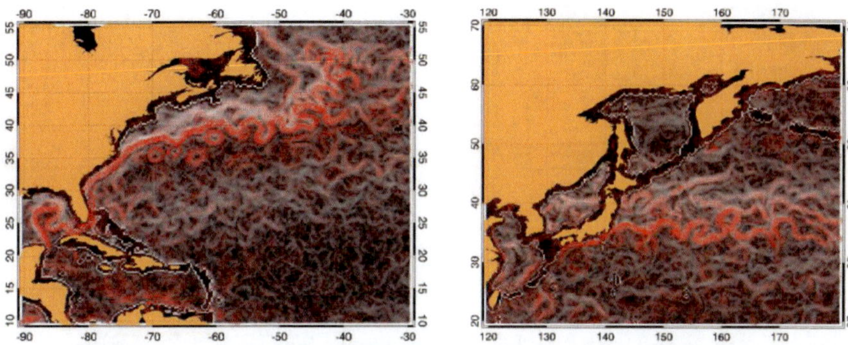

Figure 16. From top to bottom: Singularity exponents derived from MW SST for two different areas and times; altimeter-derived geostrophic velocities for the same locations and times; and superimposition of both fields; the maximum length of velocity vectors corresponds to 1 m/s. Results on the left column are for Gulf Stream area on 1 February 2003; results on the right are for Kuroshio current on 1 November 2002 (from Turiel et al., 2009).

Finally, recent work on semi-quasigeostrophic theory may open a way to retrieve the ocean velocity field (Lapeyre and Klein, 2006). From this theory, a balanced three-dimensional flow for the upper layers can be derived from the knowledge of the surface density and the characteristics of the basin scale. This can be approximately satisfied by using temperature patterns observed in single thermal images together with the aid of a climatological vertical profile of the area of interest (Isern-Fontanet et al., 2006; Isern-Fontanet et al., 2008). The results are very promising (Fig. 17) from an operational point of view because they can be automatically generated and the information required can be easily obtained and merged. Another advantage is that the method allows retrieving the vertical velocity field in such a way that can be used to provide a first guess of the vertical motions of spills.

In summary, all these well-known methodologies may provide alternative sources of ocean velocity fields that, when adequately implemented, may provide in some cases a good enough estimate of initial conditions in the region of an oil spill.

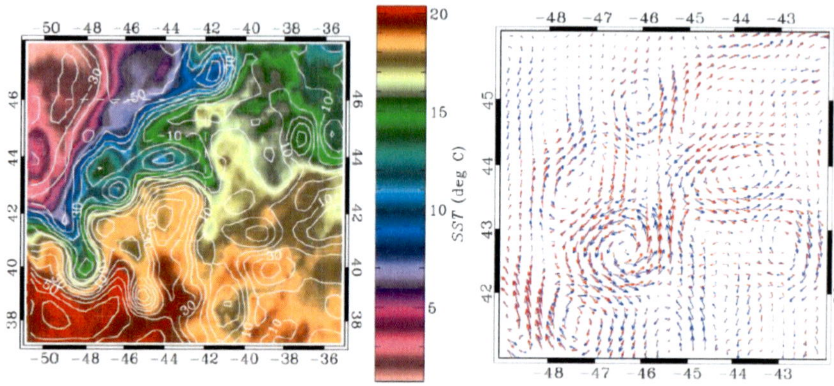

Figure 17. From left to right and from top to bottom: SST image with Absolute Dynamic Height contours superimposed (in white); velocities derived from microwave SST (red) and altimetry (blue) (band-filtered retaining scales between 100 km and 300 km) in the region delimited by the white rectangular box in the top left figure; scatter plot of the velocities located in regions with x > x0. These plots correspond to the area limited by 37N–48N, 51W–35W centered on January 5, 2005 (adapted from Isern-Fontanet et al., 2006).

4. Modeling Strategies for Oil Dispersion: Short and Long-Term Forecasts

4.1. Dispersion of Light and Heavy Fuels

Authorities are mainly concerned with: (i) economic impacts, (ii) ecological risks and (iii) peoples' security. Tools to address these three points should be designed in advance.

On the other hand, these urgent decisions have to be made with limited information. Thus, a decision tree using only expected available information in the first hours after an accident may be very useful to help authorities in an oil spill crisis.

To predict more precisely the costs and ecological risks we need to apply oil dispersion models. In the reanalysis of the Prestige accident, García-Olivares et al. (2011) used simple models which tried to predict the wind velocity field by the probability of events obtained from wind roses and modeled the ocean velocity field as a simplified mean seasonal average. However, this simple approach did not perform with enough accuracy in the long-term, since its prediction of beaching extended over 1 year instead of the observed 2 months. The use of hindcasts produced by Operational Oceanography (OO) systems provided much more accurate results.

Thus, the first step of modeling should use oil spill forecasting systems, based on OO systems, to know whether the spill will reach the coast within a few days.

If the answer is NO, the crisis may last during the entire time that there is persistence on surface (*T*) of the oil spilt. In this case, dispersion algorithms should be used to obtain the probability of beaching during the period *T*. This implies the need to obtain stochastic estimations.

For a quick estimation of long-term beaching, the concept of oil persistence on the surface can be useful. The idea is that different kinds of oils have a different persistence on the surface depending on its density. Table 1 shows four kinds of oil according to density. The ITOPF estimates average persistence for four kinds of oil, as the following figure shows.

Table 1. Different kinds of oil according to density. API degrees are obtained from the relative density (specific gravity SG) of the oil by the following expression: °API = (141.5 / SG) - 131.5

Oil Group	Density	API degrees	Examples
Group I	less than 0.8	55	Gasoline, Kerosene
Group II	0.8 - 0.85	40	Gas Oil, Abu Dhabi Crude
Group III	0.85 - 0.95	25	Arabian Light Crude, North Sea Crude Oils (e.g., Forties)
Group IV	greater than 0.95	10	Heavy Fuel, Venezuelan Crude Oils

In the case of a spill of fuel belonging to the group 1 or 2 (see Figure 18), persistence times are of the order of one week or lower. In these cases, meteorological data will be normally available for the whole integration period and should be assimilated and used by the dispersion model in order to predict the slick drift over the period. Oil spills forecasting systems based on OO systems use this approach and, when they are available, which is the usual case in developed countries, it is recommended to use them, as a first step to predict whether most of the spill will impact the coast in less than one week (the typical predictability horizon of an initial conditions simulation with an assimilation of wind data).

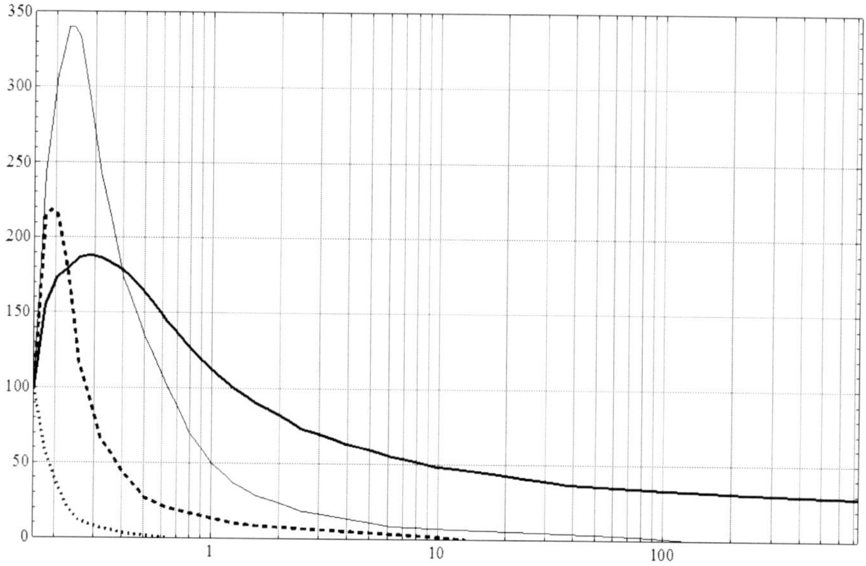

Figure 18. Percentage of persistence of oil on sea surface as a function of time (days) according to the ITOPF. Adapted from http://www.itopf.com/marine-spills/fate/models/index.html.

Surface oceanic currents also have a predictability horizon limited to several days due to their strong coupling with wind fields. In the case of more persistent kind of spills, like the spills belonging to Group 4, no forecast is available covering the period of interest, between months and years. In these cases, oil spill forecasting systems based on OO should be used, if they are available, to predict whether an important fraction of the spill will reach coastal areas within a few days after the accident. For longer time scales, these same models can be used but in the framework of statistical modeling of long-term dispersion. In these models, wind and current fields have to be probabilistically represented. OO systems can be fruitfully used in the simulation, since (i) in these models the characterization of the turbulent velocity field is state-of-the-art, and (ii) the long-term state of the ocean can be probabilistically estimated by using hindcasts obtained assuming meteorological conditions typical of the season where the spill is taking place. In addition, oil spill forecasting systems based on OO usually include models for coastal mechanisms, such as breaking waves, tides and oil interaction with sediments.

To estimate the dispersion of all kinds of oil during the first 7 days after the spill, we should run oil spill forecasting systems based on OO from the initial conditions defined by the spill.

4.2. Heavy Fuels Dispersion. The Two-Horizons Strategy

Fuels of type 3 and 4 have persistence times longer than one week. In these cases, a stochastic approach should be added to complement the initial-conditions run of oil spill forecasting systems. This stochastic approach is needed to obtain a quick estimation of long-term coastal impact (in scales longer than one week). Estimation of long-term coastal impact is needed to support the decision making in the first hours after the spill.

In particular, in many emergency situations the first dilemma is "to allow or not to allow a refuge port for the ship". In the Prestige accident, it was the first question to decide and no previous procedure was designed to answer this question. In that case, the chosen response, taking the tanker out to sea, was apparently not the best choice, as we will show in Section 5. And that wrong decision was made because no methodology was available at that moment to obtain a quick estimation of long-term beaching and its derived economic and ecologic impacts.

Results of the stochastic simulations are in general seasonally dependent. Therefore, oil spill forecasting systems should incorporate a collection of hindcasts of both wind fields and current fields organized seasonally and, possibly, considering the values of climate indexes. For example, the North Atlantic Oscillation index (NAO) has been shown to control the penetration of fronts into the North Atlantic and the Mediterranean Sea latitudes.

Stochastic simulations involve a large number of virtual particles to produce statistically significant results. Thus, it takes time (even months of calculation with a desktop computer) to be implemented. In an oil spill crisis time is a factor and, ideally, stochastic simulations should have been implemented in advance. The position of an oil spill accident cannot be predicted in advance, however the paths of high maritime traffic of tankers is known and the points on these paths have the largest probability to be sources of future oil spills.

With this in mind, in the frame of the European project MEDESS-4MS, we have obtained a set of maps relating to the beaching probability for the whole Mediterranean coast, for coastal segments of 50 km, for oil spills coming from 180 selected points on the tankers paths across the Mediterranean Sea. These paths are shown in Figure 4.

In daily hindcasts of nine years (1998/10/01 to 2007/11/01) velocity fields, which were obtained by the NEMO-MED12 oceanographic model (Fig. 19) were used. The model used 1/12° horizontal resolution stretched in latitude (varying from 6 to 8 km), 50 Z-levels surface (DZ = 1 m, near surface) down to the bottom (DZ = 460 m) with 23 levels in the first 100 m, and daily forcing at 50 km with ERA40 wind fields (ARPERA). Figure 19 shows one of these hindcasts.

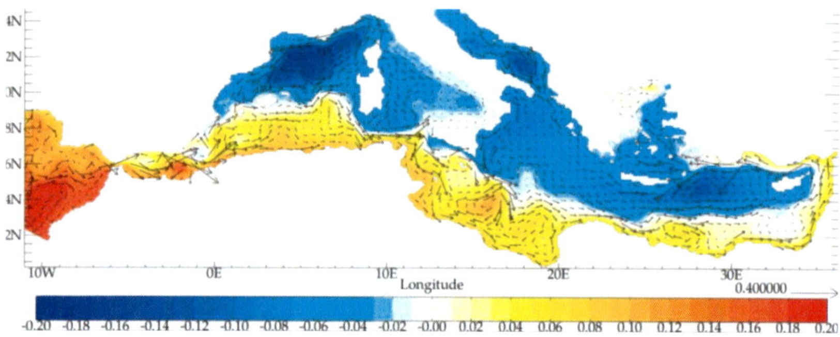

Figure 19. Snapshot of a velocity field hindcast produced by NEMO-MED12. From: Lebeaupin *Brossier* et al. (2011, 2012).

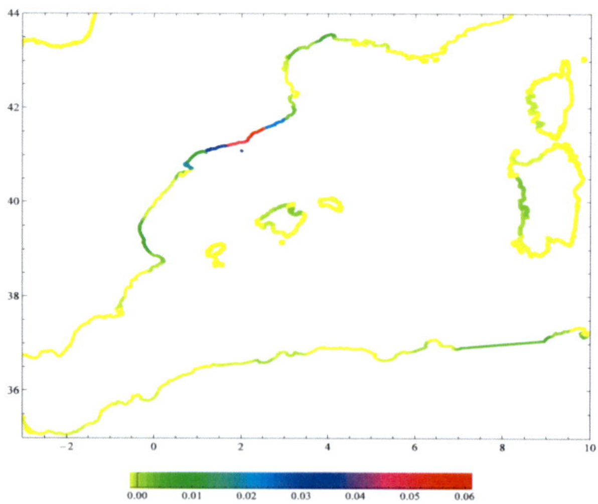

Figure 20. Beaching frequency of oil type 4 for a spill at the Casablanca oil platform (black dot in the image). The bottom bar shows the color spectrum.

Given that the persistence time of type 1 fuels is about 1 day, these fuels were not considered in our stochastic estimations of beaching. Maps of coastal beaching for oil types 2, 3 and 4 were produced for the 180 selected points. A section of one of these maps, corresponding to a spill from a point close to the Barcelona port (black dot in the image), is shown in Figure 20. KML files were generated to visualize these maps from geographical information software such as Google Earth.

The beaching time distribution (after a spill) for the most exposed segment is also obtained, as Figure 21 illustrates.

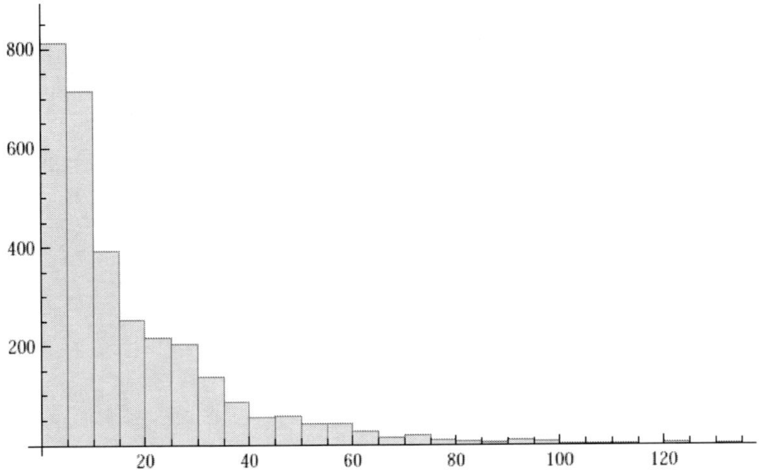

Figure 21. Beaching time distribution (after one spill at the Casablanca platform) for the most exposed segment. X-axis units are days.

These maps and distributions will be included in the Network Data Repository of the Medess-4MS system (http://www.medess4ms.eu/), where they would be available for the Decision Support System to be displayed in a user interface in case of an accident. This information could be shown in parallel with the first short-term estimations of the forecast model(s). Thus, short and long-term predictions would be present from the very beginning to the decision maker.

These sets of maps could be updated in future years with additional initial points, hindcasts longer than 10 years, statistics for different seasons and different values of the NOA index, and especially by using more sophisticated models for weathering and persistence.

4.3. Estimation of Ecological Risk

Ecological damage is not reducible in principle to monetary costs. However, it is one of the main factors to be taken into account in the evaluation of the consequences of an oil spill accident. Thus, an independent quantification of this kind of risk can be very useful in any response planning. To obtain a quick estimation of ecological risks, the following 6-step methodology may be useful (García-Olivares et al. 2011).

1. Identify the presence of ecologically sensitive targets in different coastal segments. In the case of a heavy fuel (type 4) these targets can be the following (a to h):
a) protected shorelines whose ecological dynamics should be maintained as unperturbed as possible (coastal National Parks and RAMSAR sanctuaries),
b) abundance of seabirds,
c) presence of marine mammals with coastal habitats,
d) high concentrations of molluscs with ecological value,
e) high concentrations of plankton,
f) indigenous populations of specialist sediment dwellers,
g) presence of mangrove swamps or salt marshes,
h) presence of fragile marine communities (e.g., corals, posidonia ecosystems).
2. Assign an impact value $\boldsymbol{i}$ to the hypothetical loss of the ecological value of the target, e.g.: negligible ($\boldsymbol{i}$=1), minor ($\boldsymbol{i}$=2), moderate ($\boldsymbol{i}$=3), major ($\boldsymbol{i}$=4), critical ($\boldsymbol{i}$=5) and catastrophic ($\boldsymbol{i}$=6).
3. Define a threshold value of contamination ($\boldsymbol{t}$) for which the ecological value of the target may start to be in danger. This could require detailed biological research on the species response to the oil contamination. Let us assume here that $\boldsymbol{t}$=35 t/km is the threshold for a coastal national park. This implies that, if the slick is 5 cm high and between 1 m and 5 m wide, its length along the coast would affect from 14% to 70% of the shoreline
4. Use the threshold value $\boldsymbol{t}$ to define the event "oiling (tons /km) larger than $\boldsymbol{t}$", with probability $\boldsymbol{p}$. Use the Lagrangian simulation to obtain this probability p_k(tons/km > $\boldsymbol{t}$), where k is an index to number the target.
5. Calculate $\boldsymbol{r_k} = \boldsymbol{p_k}\, \boldsymbol{i_k}$ for targets k=1,2,3...

An integrated risk could be defined from the set of r_k values obtained, for instance:

$\boldsymbol{r} = \boldsymbol{Max}(\{\boldsymbol{r}_k\})$, where ***Max*** designs the maximum of the set of values $\{\boldsymbol{r}_k\}$.

6. A range of not allowed tolerance is defined for *r*, e.g., events with $\boldsymbol{r} > \boldsymbol{1}$ are considered unacceptable. Then, the final ecological risk can be valued. The choice $\boldsymbol{r} > \boldsymbol{1}$ would imply that events that put in danger the ecological value of a target with the maximum valuation ($\boldsymbol{i}$=6 in our example) are not allowed to have a probability 1/6 or larger.

4.4. Estimation of Economic Costs

Finally, for a quick estimation of the order of magnitude of the economic costs, it is possible to use models based on historical events, e.g. Etkin (1999, 2000). The Etkin models are based on the following expressions:

$$C_e = C_u\, A\, I\, e$$

with: $C_u = C_l\, t\, o\, m\, s$, and: $C_l = r\, l\, C_n$

where the meaning of the parameters is the following:

Ce: Total Cost (Euros),
Cn: general cost per released unit,
(8600 $/tonne in Europe),
l and r: regional modifiers (1),
t: type of oil (e.g. 0.71 for heavy fuel N° 6),
o: coast length affected,
m: technology employed (e.g. for mechanical, 0.92),
s: spill size modifier,
A: Quantity spilled (tonnes),
I: Inflation rate between 1999 and the date of the accident,
e: official conversion dollars-euros.

Most of those factors have been obtained by statistical processing of the costs observed in the past accidents, and are given in tables by Etkin (1999; 2000). The length of coast affected may be obtained from the short and long-term forecasts obtained by the methodology explained in Section 4.2, and the other parameters needed by the Etkin model are expected to be known in the first hours following an oil spill accident. Thus, the use of the Etkin model is straightforward and suitable for a quick estimation of the order of magnitude of economic cost derived from coastal clean-up.

4.5. A Decision Tree for Oil Spill Emergencies

The methodology proposed here take us to the following decision tree, which can be useful to guide the decision-making about boat towing and spill control in the first moments following an oil spill crisis (Diagram 2).

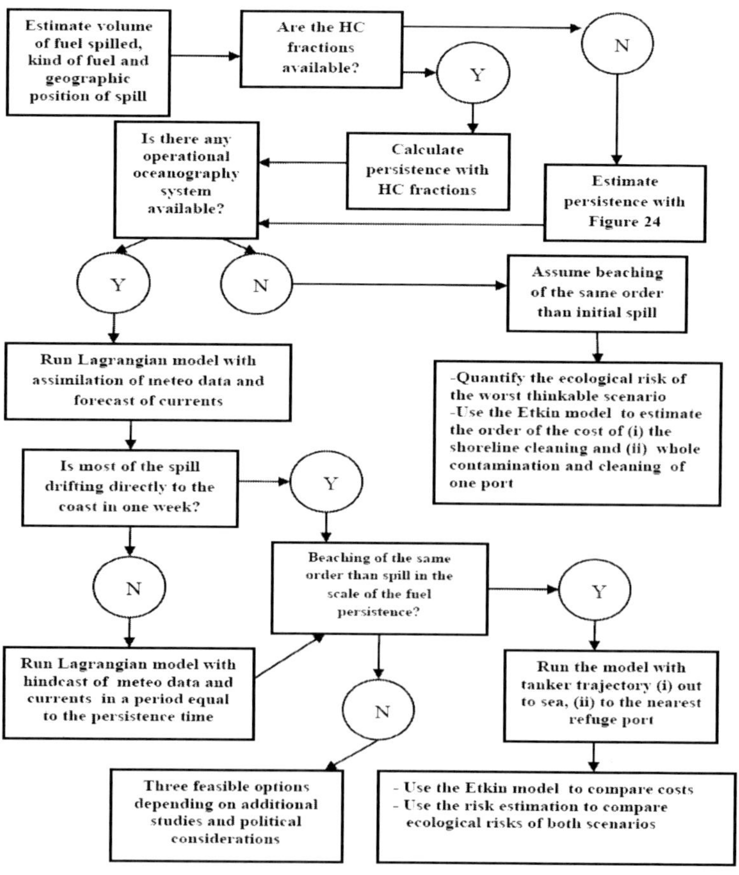

Diagram 2. Decision tree for oil spill emergencies. "HC" stands for "hydrocarbons".

5. Case Study

To illustrate the methodology proposed, we have applied it to the Prestige accident (see García-Olivares et al. 2011 for details), which was summarized in Section 2.1.

To implement the stochastic approach, hindcasts produced by the Regional Ocean Modeling System (ROMS) were used, with a mean horizontal resolution of 6.6 km. Surface forcing inputs used in the model are hourly data with a resolution of 30 km, provided by MeteoGalicia (the Regional Meteorological Office of Galicia) and obtained from the Fifth-Generation NCAR/Penn State Mesoscale Model (MM5). Hindcasts of the years 2007 and 2008 obtained with this model were used then to estimate the drift of a set of numerical Lagrangian tracers from the position of the wreckage, during a period of 365 days. One thousand particles were released every 5 days over a period of one year, to be advected during that year by the velocity fields obtained from the hindcasts. The number of drifters remaining in movement is scaled with a persistence factor to obtain the fraction of the initial spill that remains active. Two models were used to calculate the persistence factor. The first (optimistic) uses the simple curves of Fig. 18; the second (conservative) uses the content of heavy hydrocarbons in the oil to calculate the persistent fraction, which was estimated to be 70% of the fuel. The number of drifters beached over the period can be used to estimate the volume fraction that affects the shoreline. To estimate the shoreline length affected, the slick is assumed to be 5 cm high and that it forms a layer of 1 m of width when beached. With these assumptions, the slick area corresponding to any drifter may be translated to a shoreline length affected to feed the Etkin (1999) model. We emphasize that the objective is to quickly estimate the order of magnitude of the impact, and not a precise estimation of the actual impact.

Figure 22 shows the distribution of percentages of the initial volume beached in the one year period for the conservative persistence hypothesis when the hindcast approach is used for the current and wind annual fields. The frequency of no beaching events (12% in the hindcast approach; 4% in the simple approach) is not shown in the histograms. The last frequency class includes events with the maximum beaching fraction: 49% of the initial volume beached. Given that 70% of the total oil volume was not recovered (30% of the oil was assumed to be recovered by cleaning ships), 49% of the beached volume implies an effective persistence of 70%.

In the hindcast prediction the distribution is bimodal: most frequent events accumulate in the zero (frequency 12%) and 0-7% events (frequency 13%) and in the impact events between 28% and 42% of the initial spill (frequency 20 + 30 = 50%).

The initial spill from the Prestige was estimated in the range 48,904 to 61,700 t, and we can use 19,701 t as an estimation of the total oil recovered onshore, which is equivalent to 32% to 40% of the initial spill. Thus, the real beaching observed in the Prestige crisis can be considered an event belonging to the main peak of the bimodal distribution displayed in Figure 22 for the hindcast prediction.

Taking the hindcast prediction in Figure 22 as a reference, the affected shoreline length ranges between 0 and 48 km for the first set of events, with a probability of 0.25, and between 193 and 290 km for the second set of events, with a probability of 0.5.

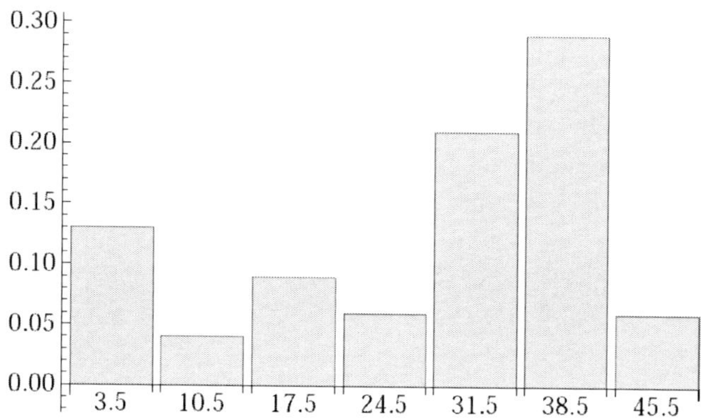

Figure 22. Distribution of percentages of the initial volume beached for the conservative persistence hypothesis when the hindcast approach is used.

Then we quantified the ecological risk associated to the presence of protected areas. Table 2 shows the predictions of our model for the different frequencies of tonnes of oil per kilometer beached in the areas surrounding the two protected areas considered. These frequencies have been calculated for releases coming from three zones close to the wreckage of the Prestige. In order to obtain enough statistics a target segment of 50 km was chosen including the Atlantic Island National Park and the Corrubedo Dunes Ramsar Park in Coruña, and another 50 km segment was centered in the "Marsh of Santoña" Ramsar Park (Cantabria). Table 3 shows the values obtained for ***p***, ***i*** and ***r*** when the methodology described in the previous section is applied. As can be observed, with the tolerance adopted in Section 4.3 ($r \leq 1$), the risk of impact on the coastal parks would be considered unacceptable in this example.

Table 2. Frequencies of different beaching classes for the Atlantic Islands National Park coast and for the "Marsh of Santoña" Ramsar Park, when the particles are released from Zones 1, 2 and 3, respectively

Atlantic Beaching (t / km)	Frequency Zone 1	Frequency Zone 2	Frequency Zone 3
0 – 3.5	0.02	0.042	0.10
3.5 - 35	0.07	0.14	0.19
35 - 85	0.06	0.12	0.14
85 - 130	0.08	0.03	0.02
130 – 173	0.06	0.03	0
173 - 216	0.02	0.01	0.01
216 - 259	0.03	0	0
259 - 302	0.02	0	0
Cantabria Beaching (t / km)			
0 – 3.5	0.002	0.028	0.03
3.5 - 35	0.008	0.028	0.08
35 - 65	0.01	0.02	0.04
65 - 98	0.00	0.02	0.08
98 - 130	0	0.01	0.02
130 - 163	0	0	0.01
163 - 195	0	0	0
195 - 228	0	0.02	0.02

Table 3. Estimation of the ecological risk for the Atlantic Islands National Park and the "Marsh of Santoña" Ramsar Park

Atlantic Islands National Park	Zone 1	Zone 2	Zone 3
p_1 ($t/km > 35$)	0.27	0.19	0.17
i_1	6 *Catastrophic*	6 *catastrophic*	6 *catastrophic*
$r_1 = p_1 \, x \, i_1$	1.62	1.14	1.02
Marsh Santoña Ramstar Park			
p_2 ($t/km > 35$)	0.011	0.07	0.17
i_2	6 *Catastrophic*	6 *catastrophic*	6 *catastrophic*
$r_2 = p_2 \, x \, i_2$	0.07	0.42	1.02
$r=Max(\{r_1,r_2\})$	1.62 *Unacceptable*	1.14 *unacceptable*	1.02 *unacceptable*

On the other hand, the Etkin model was used to estimate the order of magnitude of the economic cost derived from the clean-up of the coasts affected by the long-term beaching of oil in two different scenarios (García-Olivares et al. 2011): (i) long-term beaching derived from taking the Prestige out to sea, and (ii) short-term beaching derived from taking the Prestige to a port of refuge.

The analysis showed that the order of magnitude of the cost expected for the clean-up of the coasts in the first scenario is two orders of magnitude larger than the cost expected from the clean-up of a refuge port and its surrounding coast.

In conclusion, taking the Prestige to a port of refuge would have been a much more convenient decision than leading it out to sea, not only from an ecological perspective but also from an economic point of view.

Acknowledgments

This work has received support from the MEDESS-4M (MED 2S-MED11-01) and the TOSCA (G-MED09-425) projects funded by the MED program of the European Commission.

References

Abascal A., Castanedo S., Gutierrez A.D., Medina R., Losada I. and E. Comerma E, 2008: Description and application of the operational oil spill forecast system TESEO. Proceedings of the International Oil Spill Conference. 1023-1027.

Abascal A., Castanedo S., Medina R., Losada I. and E. Álvarez-Fanjul, 2009: Application of HF Radar currents to oil spill modeling. Marine Pollution Bulletin, 58, 238-248.

Abascal A., Castanedo S., Mendez F. J., Medina R. and Losada I. 2009. Environmental Hydraulics Calibration of a Lagrangian Transport Model Using Drifting Buoys Deployed during the Prestige Oil Spill. Journal of Coastal Research, 25, 1, 80-90.

Abascal A., Castanedo S., Medina R., and M. Liste, 2010. Analysis of the reliability of a statistical oil spill response model. Marine Pollution Bulletin, 60, 11, 2099-2110.

Al-Rabeh A.H., Lardner R.W. and N. Gunay, 2000. Gulfspill Version 2.0: a software package for oil spills in the Arabian Gulf, Environmental Modelling & Software, 15, 425–442.

Alvarez-Fanjul, E., Losada, I., Tintoré, J., Menéndez, J., Espino, M., Parrilla, G., Martinez, I., Muñuzuri, V.P., 2007. The ESEOO Project: developments and perspectives for operational oceanography at Spain. Proceedings, ISOPE-2007: the 17th International Offshore Ocean and Polar Engineering Conference. Lisbon, Portugal. The International Society of Offshore Ocean and Polar Engineers (ISOPE) 3, 1708–1715.

Balseiro C. F., Carracedo P., Gómez B., Chambel P., Naranjo L., Penabad E. and V. Pérez-Muñuzuri, 2003. Tracking the 'Prestige' Oil Spill. An operational Experience in Simulation at MeteoGalicia. Weather, 58(12), 452-458.

Beegle-Krause, C.J., 2001. General NOAA Oil Spill environment (GNOME): A new spill trajectory model. Proceedings of the 2001 International Oil Spill Conference.

Beegle-Krause, C.J., Callahan J. and C. O'Connor, 2003. GNOME extended to use nowcast/forecast currents. Proceedings of the 2003 International Oil Spill Conference.

Beegle-Krause, C.J., 2008. GNOME: NOAA's next-generation spill trajectory model. Proceedings of the 2008 International Oil Spill Conference.

Buesseler, K. O., Jayne, S. R., Fisher, N. S., Rypina, I. I., Baumann, H., Baumann, Z., Breier, C. F., Douglass, E. M., George, J., Macdonald, A. M., Miyamoto, H., Nishikawa, J., Pike, S. M. and S. Yoshida. 2011. Fukushima-derived radionuclides in the ocean and biota off Japan. Proceedings of the National Academy of Sciences of the United States of America, 109, 16, 5984–5988, doi: 10.1073/pnas.1120794109

Burns G., Ponda, R. and P. Tebeau, 2002. Looking to the Future—Setting the Agenda for Oil Spill Prevention, Preparedness and Response in the 21st Century. Spill Science & Technology Bulletin, 7, 1–2, 31–37

Camilli,R., Reddy,Ch., Yoerger D.R., Van Mooy, B., Jakuba M.,Kinsey, J.C., McIntyre C., Sylva S., Maloney J., Tracking Hydrocarbon Plume Transport and Biodegradation at Deepwater Horizon.Science, 330, 201 (2010).

Carracedo P., Torres-López S., Balseiro C. F., Montero P., Penabad E., Gómez B., Chambel P. and V. Pérez-Muñuzuri, 2006. Improvement of pollutant drift forecast system applied to the Prestige oil spills in Galicia Coast (NW of Spain): Development of an operational system. Marine Pollution Bulletin, 350-360.

Castanedo, S., Medina R., Losada, I.J., Vidal, C., Méndez, F.J., Osorio, A., Juanes, J.A. and A. Puente, 2006. The Prestige oil spill in Cantabria (Bay of Biscay). Part I: Operational forecasting system for quick response, risk assessment and protection of natural resources, Journal of Coastal Research, Vol. 22(6), 1474-1489.

CEDRE, 2002. Naufrage du Petrolier Prestige, Le lettre du Cedre, No.90, available at http://www.le-cedre.fr/images/02nov.pdf, 2002.

Coppini G., De Dominicis M, Zodiatis G, Lardner R., Pinardi N., Santoleri R., Colella S., Bignami F., Hayes D.R., Soloviev D., Georgiou G., and G. Kallos, 2011. Hindcast of oil-spill pollution during the Lebanon crisis in the Eastern Mediterranean, July–August 2006. Marine Pollution Bulletin,62, 140-153.

Crocker R.I., D. K . Matthews, W. J. Emery and D. Baldwin, 2007: Computing Coastal Ocean Surface Currents From Infrared and Ocean Color Satellite Imagery. Geoscience and Remote Sensing 45, 435-447.

Daniel P., 1997. Forecasting oil spill drift at METEO-FRANCE. Proceedings of the International Oil Spill Conference, 990-992.

Daniel P., Marty F. and P. Josse, 2003. Improvement of drift calculation in MOTHY operational oil spill Prediction System. Proceedings of the International Oil Spill Conference, 6pp.

Daniel P., P. Josse, P. Dandin, J-M. Lefevre, G. Lery, F. Cabioch and V. Gouriou, 2004. Forecasting the Prestige Oil Spills, Proceedings of the Interspill 2004 conference, Trondheim (Norway).

Daniel P. and Ph. Dandin, 2005. Benefits and use of Operational oceanography systems for drift forecasts, Proceedings of the International Oil Spill Conference, 4pp.

Daniel P., P. Josse and Ph. Dandin, 2005: Further improvement of drift forecast at sea based on operational oceanography systems, Coastal Engineering VII, Modelling, Measurements, Engineering and Management of Seas and Coastal Regions, WIT Press, pp 13-22.

Davies R, A. Horváth, C. Moroney, B. Zhang and Y. Zhu, 2007: Cloud motion vectors from MISR using sub-pixel enhancements, Remote Sensing of Environment 107 (2007) 194–199.

De Dominicis M., G. Leuzzi, P. Monti, N. Pinardi, P.-M. Poulain, 2012: Eddy diffusivity derived from drifter data for dispersion model applications. Ocean Dynamics (2012) 62:1381–1398, DOI 10.1007/s10236-012-0564-2

De Dominicis, M., Pinardi, N., Zodiatis G. and R. Lardner, 2013a. MEDSLIK-II, a Lagrangian marine surface oil spill model for short-term forecasting – Part 1: Theory. Geosci. Model Dev., 6, 1851-1869, De Dominicis, M.,

Pinardi, N., Zodiatis, G., and Archetti, R., 2013b. MEDSLIK-II, a Lagrangian marine surface oil spill model for short-term forecasting – Part 2: Numerical simulations and validations, Geosci. Model Dev., 6, 1871-1888, doi:10.5194/gmd-6-1871-2013.

Delvigne G.A.L., & C. E. Sweeney, 1988, Natural dispersion of oil. Oil & Chemical Pollution, 4, 281-310. Dietrich D.E., A. Mehra, R. L. Haney, M. J. Bowman, and Y. H. Tseng, 2004: Dissipation effects in North Atlantic Ocean modeling. Geophys. Res. Lett, 31, L05302, doi:10.1029/2003GL01901.

Elliot, A.J., 1986. Shear Diffusion and the Spreading of Oil in the Surface Layers of the North Sea," Dt. Hydrogr. Zeit 39(3), 113-137. Emery, W.J. and J.S. Dewar, 1982: Mean temperature-salinity, salinity-depth and temperature-depth curves for the North-Atlantic and the North Pacific. Prog. Oceanogr., 11: 219-305.

Emery, W. J., A.C. Thomas, M.J. Collins, W.R. Crawford and D.L. Mackas, 1986: An objective method for computing advective surface velocities from sequential infrared satellite images. Journal of Geophysical Research, 91(C11), 12865–12878.

Eremev, V.N., Ivanov, L.M., Kirwan Jr., A.D., 1992: Reconstruction of oceanic flow characteristics from Quasi-Lagrangian data: 1. Approach and mathematical methods. J. Geophys. Res. 97 ŽC6., 9733–97.

Etkin D. S. 1999. Estimating Cleanup Costs For Oil Spills. Proceedings of the 1999 International Oil Spill Conference: 35-39.

Etkin D.S. 2000. Worldwide analysis of oil spill cleanup cost factors. Proceedings of the 23rd Arctic and Marine Oilspill Program Technical Seminar: 161-174.

Fay, J.A., 1971. Physical Processes in the Spread of Oil on a Water Surface, Proceedings of the Joint Conference on Prevention and Control of Oil Spills, Washington, D.C., June, Amer. Petrol. Inst., p. 463-467.

Fernandes, R., 2001. Modelação de derrames de hidrocarbonetos. Trabalho final de curso. Instituto Superior Técnico de la Univ. de Lisboa. Available at http://maretec.mohid.com/PublicData/Products/Thesis/TFC_Rodrigo Fernandes.pdf

Font J., E. García-Gorriz and J. Candela. 1998: Data quality control for vessel mounted acoustic doppler current profiler. Application for the western Mediterranean Sea.. Scientia Marina. Vol. 61,(No. 4,): pp 417-430

French McCay, D.P., and J.J. Rowe, 2004. Evaluation of Bird Impacts in Historical Oil Spill Cases Using the SIMAP Oil Spill Model. In Proceedings of the 27th Arctic and Marine Oil Spill Program (AMOP)

Technical Seminar, Emergencies Science Division, Environment Canada, Ottawa, ON, Canada, pp. 421-452.

French-McCay D., Mueller C., Payne J., Terrill E., Otero M., Kim S.Y. Carter M., Nordhausen W., Lampinene M., Longval B., Schroeder M., Jayko K. and C. Ohlmann., 2008. Dispersed Oil Transport Modeling Calibrated by Field-Collected Data Measuring Fluorescein Dye Dispersion. Proceedings of the International Oil Spill Conference.

French-McCay D., 2009. State-of-the-Art and Research Needs for Oil Spill Impact Assessment Modeling. In Proceedings of the 32nd AMOP Technical Seminar on Environmental Contamination and Response, Emergencies Science Division, Environment Canada, Ottawa, ON, Canada, pp. 601-653, 2009.

Garcia C. A. and I.S. Robinson, 1989: Sea Surface Velocities in Shallow Seas Extracted From Sequential Coastal Zone Color Scanner Satellite Data. Journal of Geophysical Research, vol. 94, NO. C9, PP. 12,681-12,691, doi:10.1029/JC094iC09p12681.

Garcia-Górriz E., J. Candela and J. Font, 2003: Near-inertial and tidal currents detected with a vessel-mounted acoustic Doppler current profiler in the western Mediterranean Sea. Journal of Geophysical Research, VOL. 108, NO. C5, 3164, doi:10.1029/2001JC001239.

García-Ladona E., Font J., Del Río E., Julià A., Salat J., Chic O., Orfila A., Álvarez A., Basterretxea G., Vizoso G., Piro O., Tintoré J., Gil M., Herrera J.L. and S. Castanedo, 2005. The use of surface drifting floats in the monitoring of oil spills. The Prestige case. Proceedings of the 19 Biennial International Oil Spill Conference (IOSC), Miami, CD-ROM: 14718A.

García-Olivares A., de Pablos J.L., Madrigal R. (2011). Sailing the Prestige out to sea. An independent analysis. Scientia Marina, Vol 75, 533-548. doi:10.3989/scimar.2011.75n3533

GNOME (General NOAA Oil Modeling Environment) User's Manual,2002, National Oceanic And Atmospheric Administration, Office of response and Restoration, Hazardous materials Response Division. Available at http://response.restoration.noaa.gov/book_shelf/711_GNOME_Manual.pdf.

González, M., Uriarte, A., Pozo, R., Collins, M., 2006. The Prestige crisis: operational oceanography applied to oil recovery, by the Basque fishing fleet. Marine Pollution Bulletin, 53, 369–374.

Hines, A. L., and Maddox, R. N. (1985). Mass transfer: Fundamentals and applications, Prentice-Hall, Englewood Cliffs, N.J., 139

Howlett, E., Jayko K., Isaji T., Anid P., Mocke G. and F. Smit, 2008. Marine forecasting and oil spill modeling in Dubai and the gulf region. Proceedings of the 31st AMOP Technical Seminar on Environmental Contamination and Response, COPEDEC VII, 2008, Dubai, UAE.

Informe sobre el accidente del Buque "Prestige", 2002. Comisión Permanente de Investigación de Siniestros Marítimos. Dirección de la Marina Mercante del Ministerio de Fomento de España. 161 pp. (available at http://www.fomento.es/NR/rdonlyres/407AF7E2-D2FC-496A-B327-5F29F8C963A1/55204/Prestige13Nov2002.pdf)

Isern-Fontanet J., B. Chapron, G. Lapeyre and P. Klein, 2006: Potential use of microwave Sea Surface Temperatures for the estimation of ocean currents. Geophys. Res. Lett., 33, L24608, doi:10.1029/2006GL027801.

Isern-Fontanet J., G. Lapeyre, P. Klein, B. Chapron, M. Hecht, 2008: Three-dimensional reconstruction of oceanic mesoscale currents from surface information. J. Geophys. Res., 113, C09005, doi:10.1029/2007JC004692.

ISM-Code: International Safety Management Code. International Maritime Organization (IMO) available at http://www.imo.org/OurWork/HumanElement/SafetyManagement/Pages/ISMCode.aspx

ITOPF, 2013, Oil Tanker Spill Statistics 2013, The International Tanker Owners Pollution Federation Limited, available at http://www.itopf.com/information%2Dservices/publications/documents/OilSpillstats_2013.pdf

Johannessen J.A., B. Chapron, F. Collard, F. Kudryavtsev, A. Mouche, D. Akimov and K.-F. Dagestad, 2008: Direct ocean surface velocity measurements from space: Improved quantitative interpretation of Envisat ASAR observations. Geophysical Research Letters, 35 (22): Pages 1-6.

Jordà G., Comerma E., Bolaños R., Espino M., 2007, Impact of forcing errors in the CAMCAT oil spill forecasting system. A sensitivity study, Journal of Marine Systems, 65, 1–4, 134–157.

Lapeyre G. and P. Klein, 2006: Dynamics of the upper oceanic layers in terms of surface quasigeostrophy theory. Journal of Physical Oceanography, 36, 165-176 (2006).

Lardner, R.W., Zodiatis, G., 1998. An operational oil spill model in the Levantine Basin (Eastern Mediterranean Sea). Int. Symp. Mar. Pollut. 10, 5–9.

Lardner, R., Zodiatis, G., Hayes, D. and N. Pinardi, 2006. Application of the MEDSLIK Oil Spill Model to the Lebanese Spill of July 2006. European Group of Experts on Satellite Monitoring of Sea Based Oil Pollution. European Communities.

Lebeaupin Brossier C., K. Béranger, C. Deltel and P. Drobinski. The Mediterranean response to different space-time resolution atmospheric forcings using perpetual mode sensitivity simulations. In /Ocean Modelling/, Vol. 36(1-2) 1-25, DOI: 10.1016/j.ocemod.2010.10.008), 2011

Lebeaupin Brossier C., K. Béranger and P. Drobinski. Sensitivity of the northwestern Mediterranean Sea coastal and thermohaline circulations simulated by the 1/12 degrees-resolution ocean model NEMO-MED12 to the spatial and temporal resolution of atmospheric forcing. In /Ocean Modelling/, Vol. 43-44, 94-107, DOI: 10.1016/j.ocemod.2011.12.007), 2012

Liu Y, Weisberg R., Hu Ch. and L. Zheng, 2011, Tracking the Deepwater Horizon Oil Spill: A modeling perspective. EOS, 92, 5, pp:45-46.

Machín F., M. Emelianov, P. Rodríguez, E. García-Ladona, J. Menéndez, J. Salat, 2008: XBT profilers for operational purposes: application and validation in real exercises. Scientia Marina, 72, 4, 779-799. DOI: 10.3989/scimar.2008.72n4779.

Mackay, D., Leinonen, P.J., 1977. Mathematical model of the behaviour of oil spills on water with natural and chemical dispersion. Tech. Review Report no. EPS-3-EC-77-19, Fisheries and Environment Canada.

Mackay, D., Paterson, S., 1980. Calculation of the evaporation rate of volatile liquids. Proc. 1980 National Conf. on Control of Hazardous Material Spills, Louiville, Ky \

Mancho, A. M., Small, D. & Wiggins, S. 2006, A comparison of methods for interpolating chaotic flows from discrete velocity data. Comp. & Fluids 35, 416 428.

MARPOL: International Convention for the Prevention of Pollution from Ships. International Maritime Organization (IMO), available at http://www.imo.org/About/Conventions/ListOfConventions/Pages/International-Convention-for-the-Prevention-of-Pollution-from-Ships-%28MARPOL%29.aspx

Miranda, F., Braunschweig, F., Leitao, P., Neves, R., Martins, F. and A. Santos, 2000. MOHID 2000, a coastal integrated object oriented model. In: Hydraulic Engineering Software VIII. WIT Press.

Montero P., Blanco J., Cabanas J.M., Maneiro J., Pazos Y., Moroño A., Balseiro C.F., Carracedo P., Gómez B., Penabad E., Perez-Muñuzuri V., Braunschweig F., Fernades R., Leitao P.C., Neves R. 2003. Oil Spill Monitoring, Forecasting on the Prestige-Nassau accident. in: Proceedings

26th Artic, Marine Oil Spill Program (AMOP) Technical Seminar 2: 1013–1029. Environment Canada, Ottawa, Canada.

National Commission on the BP Deepwater Horizon Oil Spill and Offshore Drilling, 2011: Deep Water: The Gulf Oil Disaster and the future of offshore drilling. Report to the President. Available at http://www.gpo.gov/fdsys/pkg/GPO-OILCOMMISSION/pdf/GPO-OILCOMMISSION.pdf

Notarstefano P., P.M. Poulain and E. Mauri, 2008: Estimation of Surface Currents in the Adriatic Sea from Sequential Infrared Satellite Images. J. Atmos. Ocean. Tech., vol. 25, 271-284.

Okubo, A. (1971). Oceanic diffusion diagrams. Deep-Sea Research 18, 789-802.

OPA: Oil Pollution Act. United States Environmental Protection Agency (EPA), available at http://www.epa.gov/oem/content/lawsregs/opaover.htm.

Pascual, A., Faugère, Y., Larnicol, G and PY. Le Traon, 2006, Improved description of the ocean mesoscale variability by combining four satellite altimeters. Geophysical Research Letters 33 (2).

Payne, J.R., B.E. Kirstein, G.D. McNabb, Jr., J.L. Lambach, R. Redding, R.E. Jordan, W. Hom, C. deOliveira, G.S. Smith, D.M. Baxter, and R. Geagel. Multivariate Analysis of Petroleum Weathering in the Marine Environment – Subarctic. U.S. Department of Commerce, National Oceanic and Atmospheric Administration, OCSEAP Vol. 21 and 22, 690p, 1984

Poulain P-M. and P.P. Niiler, 1989: Statistical analysis of the surface circulation in the California Current System using satellite-tracked drifters. J Phys Oceanogr 19(10):1588–1603.

Press W.H., Teukolsky S.A., Vetterling W.T., Flannery B.P. , 1996, Numerical Recipes in Fortran 90: The Art of Parallel Scientific Computing, Volume 2 of Fortran Numerical Recipes, Second Edition (New York: Cambridge University Press, 1996)

Reed, M., Turner, C., Odulo, A., Isaji, T., Sorstrom, S. & Mathisen, J. P. (1990). Field evaluation of satellite-tracked surface drifting buoys in simulating the movement of spilled oil in the marine environment. Final Report to U.S. Dept. Interior, Minerals Management Service, Contract No. 14-35-0001-30485, MMS Tech. Report 90-0050, 213 pp.

Reed, M., Turner, C., Odulo, A., 1994. The role of wind and emulsification in modelling oil spill and drifter trajectories. Spill Science and Technology Bulletin 1 (2), 143-157.

Reed, M., Aamo, O.M., Daling, P.S., 1995. Quantitative analysis of alternate oil spill response strategies using OSCAR. Spill Science and Technology Bulletin 2 (1),67-74.

Reed, M., Johansen E., Brandvik P.J., Daling P., Lewis A., Fiocco R., Mackay D. and R. Prentki, 1999. Oil Spill Modeling towards the Close of the 20th Century: Overview of the State of the Art. Spill Science & Technology Bulletin, Vol. 5, No. 1, pp 3-16,

REMPEC, 2011: IMO/UNEP: Regional Information System; Part C2, Statistical Analysis - Alerts and Accidents Database, 2011. Regional Marine Pollution Emergency Response Centre for the Mediterranean Sea (REMPEC, www.rempec.org).

Ruiz S., J. Font, G. Griffiths and A. Castellón, 2002: Estimation of heading gyrocompass error using a GPS 3DF system: Impact on ADCP measurements Sci. Mar., 66 (4): 347-354.

Sayol J.M., Orfila J.M., Simarro G., Conti D., Renault L. and A. Molcard, 2014, A Lagrangian model for tracking surface spills and SaR operations in the ocean, Environmental Modelling & Software, 52, 74–82.

Scholz D.K. Kucklick, J.H, Pond R., Walker, A.H., Bostrom A and P. Fischbeck, 1999, Fate of spilled oil in marine waters: Where Does it go ?, What does it do ?, How do Dispersants affect it ?.An information Booklet for Decision-Makers, American Petroleum Institute, nº 4691.

Sebastião, P. and Guedes Soares, C., 1995, Modelling the fate of oil spills at sea. Spill Science Technology Bulletin, 121.

Sotillo M.G., Alvarez Fanjul E., Castanedo S., Abascal A.J., Menendez J., Emelianov M., Olivella R., García-Ladona E., Ruiz-Villarreal M., Conde J., Gómez M., Conde P., Gutierrez A.D. and R. Medina. 2008. Towards an operational system for oil-spill forecast over Spanish waters: Initial developments and implementation, Marine Pollution Bulletin, 56, 686–703.

Spaulding M.L., Howlett E., Anderson E.L. and K. Jayko. 1992. OILMAP: A global approach to spill modeling. Proceedings, 15th Arctic Marine Oil Spill Program (AMOP) Technical Seminar, Edmonton, AL, Canada, June 9-11, 1992, Emergencies Science Division, Environment Canada, Ottawa, ON, Canada, pp 15-21.

Spaulding M.L., Kolluru V., Anderson E. and Howlett. 1994. Application of three dimensional oil spill model (WOSM/OILMAP) to hindcast the Braer Spill. Spill Science and Technology Bulletin 1:23-35.

Thacker, W.C and L. Sindlinger, 2006: Estimating salinity to complement observed temperature: 2. Northwestern Atlantic. Journal of Marine Systems, doi:10.1016/j.marsys.2005.06.007.

Tkalich P. and Xiaobo C. (2001). ACCURATE SIMULATION OF OIL SLICKS. International Oil Spill Conference Proceedings: March 2001, Vol. 2001, No. 2, pp. 1133-1137. doi: http://dx.doi.org/10.7901/2169-3358-2001-2-1133

Turiel, A., J. Isern-Fontanet, E. García-Ladona and J. Font, 2005: Multifractal method for the instantaneous evaluation of the stream function in geophysical flows, Phys. Rev. Lett., 95, 104502, doi:10.1103/ PhysRevLett.95.104502.

Turiel A., J. Solé, V. Nieves, J. Ballabrera-Poy and E. García-Ladona, 2008: Tracking oceanic currents by singularity analysis of Micro-Wave Sea Surface Temperature images, Rem. Sens. Environ., 112, 2246–2260.

Turiel A., Nieves V., Garcia-Ladona E., Font J., Rio M.H., and G. Larnicol, 2009: The multifractal structure of satellite sea surface temperature maps can be used to obtain global maps of streamlines. Ocean Sci., 5, 447–460.

Wang J., & Y. Shen. Modeling oil spills transportation in seas based on unstructured grid, finite-volume, wave-ocean model. Ocean Modelling, 35, 4, 2010, pp 332–344.

White I. C. and Molloy F. C. 2003. Factors that determine the cost of oil spills, The International Tanker Owners Pollution Federation Limited (2003), London, IOSC 2003 ID#83.

Zadeh, E.S., and Hejazi, K. 2012. Eulerian Oil SpillsModel Using Finite-Volume Method with Moving Boundary and Wet-Dry Fronts, Modelling and Simulation in Engineering 2012, Article ID 398387, 7 pages. doi:10.1155/2012/398387. http://dx.doi.org/10.1155/2012/398387.

In: Marine Pollution
Editor: Dominic E. Jefferson
ISBN: 978-1-63321-520-7

Chapter 3

MARINE POLLUTION: TYPES, ENVIRONMENTAL SIGNIFICANCE AND POLLUTION CONTROL

Alberto Zafra-Gómez*, Noemí Isabel Dorival García and Inmaculada Jiménez-Díaz

Research Group of Analytical Chemistry and Life Sciences, Department of Analytical Chemistry, University of Granada, Campus of Fuentenueva, Granada, Spain

ABSTRACT

Despite the fact that water covers two-thirds of the Earth's surface, it is surprisingly vulnerable to human influence and activity. Marine ecosystems receive large amounts of a variety of pollutants from either treated or untreated wastewater. The extensive use of chemicals and their introduction into the environment has resulted in increased public concern for the potential threats to ecosystems, human health and safety. Persistent organic pollutants such as polychlorinated biphenyls, polycyclic aromatic hydrocarbons, polybrominated diphenyl ethers and pesticides, as well as heavy metals, surfactants; endocrine disrupting chemicals or personal care products and pharmaceuticals are pollutants usually present in marine environments. In recent years, as a result of this concern for environmental pollutants, there is a need to study and control

* E-mail: azafra@ugr.es.

the presence of a large amount of compounds in different environmental compartments, as many of them show carcinogenic, mutagenic or endocrine activity on living organisms. Wastewater discharges to the sea are the main route for these compounds to reach marine environments. In addition, the cumulative effects of these pollutants on coastal environments are expected to be considerable. Pollutants are always present in sediments or waters and from where they can reach animals and plants. Phyto and zooplankton absorb and accumulate toxic chemicals. Plankton is eaten by small fish that are, in turn, eaten by larger fish. As consequence, the toxic compound enters the food chain in high concentrations due to the effect of biological magnification. Over the last few years, legislation has limited the use of many chemicals because of their potential to produce adverse effects on wildlife and ecosystems. In this context, it is necessary the development and validation of analytical methods for the detection of these compounds in different marine compartments, which allows us to know the fate of these chemicals, in order to prevent, limit or reduce the impact of environmental contamination. The aim of the present work is to provide a comprehensive overview of the recent developments related to the determination of pollutants in the different marine compartments (water, sediments and biota), with special emphasis on sample preparation and analytical techniques as well as the achieved detection limits.

LIST OF ACRONYMS AND ABBREVIATIONS

4-MBC	4-methylbenzylidene camphor
2,4-DCP	2,4-dichlorophenol
2,4,6-TCP	2,4,6-trichlorophenol
ACN	Acetonitrile
Ace	Acetone
AEO	Alcohol polyethoxylate
AES	Alcohol ethoxy sulphate
AP	Alkylphenol
APCI	Atmospheric pressure chemical ionization
APEC	Alkylphenoxy carboxylate
APEO	Alkylphenol ethoxylate
AS	Alcohol sulphate
BDE	Brominated diphenyl ether
BP	Benzophenone
BP-3	Benzophenone-3
BPA	Bisphenol A

BrMMC	4–bromomethyl–7–methoxycoumarin
BSTFA	N,O–bis(trimethylsilyl)trifluoro ethyl acetamide
BTBPE	1,2–bis(2,4,6– tribromophenoxy)ethane
CDEA	Coconut diethanol amide
CE	Capillary electrophoresis
DAD	Diode array detector
DBDPE	Decabromodiphenyl ethane
DCM	Dichloromethane
DDD	Dichlorodiphenyldichloroethane
DDE	Dichlorodiphenyldichloroethylene
DDT	Dichlorodiphenyltrichloroethane
DEET	N,N-diethyl-*m*-toluamide
DES	Diethylstilbestrol
DHDMB	2,2-dihydroxy-4,4-dimethoxybenzophenone
DLLME	Dispersive Liquid-Liquid Microextraction
DP	Dechlorane Plus
dw	Dried weight
E1	Estrone
E2	17β-estradiol
E3	Estriol
ECD	Electron capture detector
ECNI	Electron capture negative ionization
EDC	Endocrine disrupting chemical
EDTA	Ethylenediaminetetraacetic acid
EE2	17α-ethinylestradiol
EHMC	Ethylhexyl methoxy cinnamate
EHS	Ethylhexyl salicylate
EI	Electron impact
ELISA	Enzyme-Linked ImmunoSorbent Assay
ESI	Electrospray ionization
EtAc	Ethyl acetate
EtOH	Ethanol
EU	European Union
FAE	Fatty alcohol ethoxylate
FMASE	Focused microwave assisted soxhlet extraction
GC-ECD	Gas chromatography electron capture detector
GC-MS	Gas chromatography-Mass spectrometry
GPC	Gel permeation chromatography
HBCD	Hexabromocyclododecane

HCB	Hexachlorobenzene
HCH	Hexachlorocyclohexane
HD–SPME	Hydro distillation solid–phase microextraction
Hex	Hexane
HLB	Divinylbenzene/*N*-vinylpyrrolidone
HPLC	High performance liquid chromatography
HS–SPME	Headspace solid–phase microextraction
HTAB	Hexadecyltrimethylammonium bromide
IDL	Instrumental detection limit
IT	Ion trap
ITMS	Ion trap mass spectrometry
LAS	Linear alkylphenol sulfonates
LC-FLD	Liquid chromatography with fluorescence detection
LC-MS	Liquid chromatography with mass spectrometry detection
LLE	Liquid–liquid extraction
LOD	Limit of detection
LOQ	Limit of quantification
LRMS	Low resolution mass spectrometry
lw	Lipid weigth
MAE	Microwave–assisted extraction
MAME	Microwave assisted micellar extraction
MASE	Membrane assisted solvent extraction
MDPP	Methyl diphenyl phosphate
MeOH	Methanol
MISPE	Molecularly imprinted solid–phase extraction
MSPD	Matrix solid phase dispersion
MTBSTFA	N–(terbutyldimethylsilyl)–N–methyl trifluoroacetamide
NBFR	Novel brominated flame retardants
NCI	Negative chemical ionization
NH_4Ac	Ammonium acetate
NP	Nonylphenol
NPEO	Nonylphenol polyethoxylate
NSAID	Non-steroidal anti-inflammatory drug
OC	Octocrylene
OCP	Organochlorine pesticide
OP	Octylphenol
OPFR	Organophosphate flame retardant
PAH	Polycyclic aromatic hydrocarbon
PBCCH	Pentabromochorocyclohexane

PBDE	Polybrominated diphenyl ether
PBB	Polybrominated biphenyl
PBs	Paraben
PBT	2,3,4,5,6–pentabromotoluene
PCB	Polychlorinated biphenyl
PCDD	polychlorinated dibenzo-dioxin
PCDF	Polychlorinated dibenzo-furan
PC-HFME	Polymer-coated hollow fiber microextraction
PCP	Personal care product
PEG	Polyethylene glycol
PFBCI	Pentafluorobenzoylchloride
PFBS	Perfluorobutanesulfonate
PFC	Perfluorinated compound
PFDA	Perfluorodecanoic acid
PFDoDA	Perfluorododecanoic acid
PFDS	Perfluorodecanesulfonate
PFHpA	Perfluoroheptanoic acid
PFHS	Perfluorohexanesulfonate
PFHxA	Perfluorohexanoate
PFHxS	Perfluorohexanesulfonate
PFNA	Perfluorononanoic acid
PFOA	Perfluorooctanoate
PFOS	Perfluorooctanesulfonate
PFOSA	Perfluorooctane sulfonamide
PFUnDA	Perfluoroundecanoic acid
PLE	Pressurized liquid extraction
POLE	Poly oxyethylene lauryl ether
POP	Persistent organic pollutant
PPCP	Pharmaceuticals and personal care product
QAC	Quaternary ammonium compound
QToF	Quadrupole time of flight
SAESC	Sonication–assisted extraction in small columns
SBSE	Stir bar sorptive extraction
SDME	Single drop microextraction
SEC	Size exclusion chromatography
SIM	Selected ion monitoring
SLE	Solid-liquid extraction
SPC	Sulphophenyl carboxylate compound
SPE	Solid phase extraction

SPME	Solid–phase microextraction
TBA	Tetrabutylammonium
TBBPA	Tetrabromobisphenol A
*t*BP	4-*tert*-butylphenol
TBPhA	Tetrabromophthalic anhydride
TBT	Tributiltin species
pTBX	2,3,5,6–tetrabromo–*p*–xylene
TCA	Trichloroacetic acid
TCBPA	Tetrachlorobisphenol A
TCEP	Tris(2-chloroethyl) phosphate
TCM	Trichloromethane
TCPP	Tris (2-chloroisopropyl) phosphate
TCS	Triclosan
TDBPP	Tris(2,3–dibromopropyl) phosphate
TDCPP	Tris(1,3-dichloro-2-propyl) phosphate
TiBP	Tri-iso-butyl phosphate
ToF	Time of flight
tOP	4-*tert*-octylphenol
TPP	Triphenyl phosphate
UPLC	Ultra performance liquid chromatography
USE	Ultrasound–assisted extraction
UV	Ultraviolet radiation

1. Introduction

Marine waters play a key role in the development of life and in numerous life cycles. Human activities are causing chemical changes in the oceans [1]. Decades of pollution from industrial and domestic sources, coastal habitat destruction, pollutant emissions, overfishing and bottom trawling have had devastating impacts on biodiversity and habitats [2, 3]. The problems caused to marine ecosystems are directly or indirectly associated to the world's growing population. It has been shown that more than 40% of the world's marine waters are highly affected by human activities [4], with the biggest impacts felt along the coastal areas [5].

For centuries, insufficient attention has been paid to the critical impacts of overfishing and algal blooms on marine ecosystems particularly by developing countries. At the same time, climate change is affecting the oceans in major ways that we are only beginning to understand. Considering the importance of a sustainable use of marine resources and biodiversity, world communities should pay much more attention to the solution of current problems created on marine ecosystems.

In recent years, as a result of this concern for environmental pollutants, there is a need to study and control the presence of a large amount of compounds in different environmental compartments, as many of them show carcinogenic, mutagenic or endocrine activity on living organisms. In addition, the cumulative effects of these pollutants on coastal environments are expected to be considerable. Pollutants are always present in sediments or waters and from where they can reach animals and plants. Phyto and zooplankton absorb and accumulate toxic chemicals.

The Joint Group of Experts on the Scientific Aspects of Marine Pollution (GESAMP), a United Nations-sponsored advisory body, has defined marine pollution as "*the introduction by man, directly, or indirectly, of substances or energy to the marine environment resulting in deleterious effects such as: hazards to human health, hindrance to marine activities, impairment of the quality of seawater for various uses and reduction of amenities*" [6]. Hence, evaluation of marine pollution is paramount not only to assess the global effects of environmental contamination but also to ensure the quality of marine resources for human consumption.

Over the last few years, legislation has limited the use of many chemicals because of their potential to produce adverse effects on wildlife and ecosystems. For example, the EU has developed an ambitious plan for the protection of the marine environment and the remediation of the affected areas by 2020. Additionally, whose aim is to protect more effectively the marine environment across Europe was adopted in 2008 [7]. As a result of this legislation, the study of the occurrence pollutants in different marine compartments have become a major threat to the health of ecosystem due to they do not break down easily, so that their residues accumulate in different species of marine organisms [8-11].

The uptake of xenobiotics by phytoplankton is the first step in the bioaccumulation in aquatic food webs. Biomagnification of these pollutants through the food chains results in dispersion of pollutants far from their sources and from areas of human activity, e.g. PCBs in deep-sea fish [11], whales [8] and seals in oceanic waters and in polar bears [9, 12] and seabirds [13]. Animals higher up in the food chain, such as seals, can have contamination levels millions of times higher than the water in which they live. And polar bears, which feed on seals, can have contamination levels up to 3 billion times higher than their environment. Thus, it is obvious that the impact of these toxic and persistent pollutants on marine ecosystems could be pronounced in terms of their biological responses in organisms [14]. It is important to remark that the toxic effects caused by many of those pollutants on aquatic organisms can appear at very low concentration levels [16-18].

Risk assessment studies based only on data derived from water or sediment analyses may be misleading. Data from sediments are not representative of contamination in the overlying water column and cannot give information on patterns of contamination at the higher levels of the food chain [19].

The study of the final fate or behavior of these chemicals in marine organisms at different levels of the food chain is essential to have a complete view of the actual effects, threats and risks of the presence of large amounts of pollutants in marine ecosystems.

In this context, the main objective of the present work is to provide a comprehensive overview of the recent developments related to the determination of pollutants in the different marine compartments (water, sediments and biota), with special emphasis on sample preparation and analytical techniques as well as the achieved detection limits. Aspects such as biomonitoring and bioremediation are also treated.

2. Marine Pollutants

The marine environment has traditionally been considered as a vast reservoir for the safe disposal of a large variety of pollutants. However, many chemical contaminants, composed of diverse groups of both organic and inorganic compounds, are now known to have potential health effects on organisms such as humans, fish and wildlife, and may have other adverse ecological effects [15], even when they are released at low levels (usually at concentration levels of nanograms to micrograms per liter) [16, 17].

The introduction of the European REACH regulation has led to the development of less persistent, bioaccumulative, or toxic chemicals [20], with different properties in comparison with the famous "dirty dozen" from the Stockholm Convention (2001), mainly characterized as hydrophilic compounds. As a consequence, the emerging more polar anthropogenic contaminants, such as pharmaceuticals, PFCs and pesticides, have recently gained more attention. Other compounds, such as PCPs, show some non polar features, establishing important interactions in non-polar matrices.

On the other hand, compounds such as PAHs, PCBs and OCPs, are still the most important types of ubiquitous priority pollutants, although many of them were banned many years ago. However, due to their persistence, these chemicals are still present in the environment and represent a concern because of their mutagenic, carcinogenic and endocrine disrupting effects on biota.

Some of the most important considered as highly persistent pollutants include PCBs, PAHs, PCDD, PCDFs, TBT, nitroaromatics, radionuclides, OCPs (DDT, HCH, aldrine, dieldrin, endrine, heptaclor, etc.), herbicides, toxic metals (Cd, Hg, Ag, Co, Cr, Ni, Pb, Zn, Cu, As, Sb, etc.), surfactants, petroleum products, the so-called “emerging contaminants” that include pharmaceuticals, PCPs, EDCs, such as steroids and hormones, PFCs, brominated flame retardants including polybrominated diphenyl ethers, sun-screens/UV filters, fragrances, antiseptics or plasticizers [15, 18, 19, 21].

2.1. Polycyclic Aromatic Hydrocarbons

PAHs are ubiquitous pollutants with well-known carcinogenic potential. PAHs originate from anthropogenic sources and natural sources [22]. They are of concern because of their persistence and potential to accumulate in aquatic organisms. PAHs include two (e.g. naphthalene), three (e.g. anthracene, fluorene, phenanthrene), four-ring (e.g. fluoranthene, pyrene, benzo(a)anthracene), and five-ring compounds (e.g. benzo(b)fluoranthene, perylene and benzo(a)pyrene).

2.2. Polychlorinated Biphenyls and Polybrominated Diphenyl Ethers

PCBs are a group of synthetic halogenated aromatic hydrocarbons consisting of a complex mixture of chlorinated biphenyls (209 possible congeners) that contain a varying number of substituted chlorine atoms on aromatic rings. PCBs are discharged into the aquatic environment mainly through effluents (principally from paper factories), or as a result of accidents (leakage from PCB-containing equipment). Due to their highly hydrophobic character, PCBs are primarily adsorbed to particles and their high persistence leads to bioaccumulation in organisms.

PBDEs are a class of brominated flame retardants. They are structurally similar to PCBs, also consisting of a complex mixture of 209 possible congeners. Human exposure to PBDEs is thought to be from household products with the subsequent release into the aquatic environment through sewage treatment plant effluents.

2.3. Organochlorine Pesticides

OCPs are defined as persistent organic pollutants by the Stockholm Convention because of their persistence, bioaccumulation and potential for long-range transport. OCPs are transported by the wind in the form of vapor, particles or droplets, and carried for long distances away from the original source. Precipitation deposits the contaminants on waterways where they may accumulate. The most studied OCPs are DDT and metabolites (DDE and DDD), α- and β-endosulfan, heptachlor and HCB/HCH isomers.

2.4. Pharmaceuticals and Personal Care Products

PPCPs refer in general to any product used by individuals for personal health or cosmetic reasons or used by agribusiness to enhance growth or health of livestock. PPCPs comprise a diverse collection of thousands of chemical substances, including prescription and over-the-counter therapeutic drugs, veterinary drugs, fragrances and cosmetics. It is well-established that sewage treatment plant effluents are the main sources of PPCPs in the aquatic environment [23]. Although their persistence is low, they are considered to be semi-persistent due to their continuous introduction. Some of the most studied PPCPs in marine environments are antibiotics, analgesic/anti-inflammatory and psychiatric drugs, surfactants and UV-filters.

Exposure to pharmaceuticals may have different adverse effects in aquatic species. PPCPs have additive acute or chronic toxicity and a large number of exposure-related health effects have been described [24-27]. As far as ecosystems are concerned, effects may include biodiversity loss at different levels. The intersex condition, in the most severely affected fish, is associated with reduced fertility [28]. It has been shown that the antidepressants fluoxetine, sertraline and their metabolites cause developmental and endocrine changes [29]. Diclofenac residues are responsible for the decline of vulture populations and for the growth inhibition of marine phytoplankton [27, 30]. Antibiotics such as tetracycline at environmental concentrations (up to mg L^{-1}) may induce development of resistance in microbial assemblages, may have adverse effect on the immune system, and may inhibit growth and reproduction in fish, microorganisms, algae and aquatic plants [25, 26, 31-33]. Antibiotics may affect fish indirectly by modulating microbial functions in aquatic ecosystems modifying processes such as denitrification, nitrogen fixation or organic breakdown [34]. Phytoplankton is more affected than zooplankton and other aquatic organisms [35] and this situation is dangerous since phytoplankton is at the base of marine food chain.

2.5. Surfactants

Surfactants are organic chemicals that reduce surface tension of liquids. There are hundreds of compounds used as surfactants and many sources of surfactants are discharged into natural waters. Industrial sources include textile and detergent formulation. Surfactants are also used in laundries and households and are therefore found in discharges from sewage treatment works. The most used and studied surfactants include LAS, AEO and AES.

2.6. Endocrine Disrupting Chemicals

EDCs cover a wide range of synthetic and natural substances able to alter the normal hormone function of wildlife and humans by mimicking or inhibiting endogenous hormone action, modulating the production of endogenous hormones, or altering hormone receptor populations. Among these substances are included OCPs, PCBs, APs, phthalates, dioxins, bisphenols and natural hormones [36]. Most EDCs in the aquatic environment originate from the following major sources: anthropogenic emissions, natural production and photochemical and/or microbial production [37, 38].

2.7. Heavy Metals

Although there is no clear definition of heavy metal, density is in most cases the defining factor. Thus, heavy metals are commonly defined as those having a specific density of more than 5 g cm^{-3}. They can be transported to the sea via rivers, run-off in coastal areas, waterborne discharges of industrial and municipal waste or wet and dry atmospheric deposition. The most important and commonly studied heavy metals are lead, chromium, cadmium, copper, zinc, mercury and arsenic as well as nickel and vanadium, which sometimes are used as markers for oil contamination [39-41].

The determination of heavy metals has traditionally been carry out through concentration techniques, including co-precipitation with $Mg(OH)_2$ [40], solvent extraction [41], and the use of a chelating resin [42]. On the other hand, the most commonly used analytical methods have been flame atomic absorption spectrometry [43], electrothermal atomic absorption spectrometry [44], inductively coupled plasma optical emission spectrometry [45], and inductively coupled plasma-mass spectrometry [46].

3. Analytical Methods for the Analysis of Pollutants in Seawater and Marine Sediments

3.1. Sample Treatment: Pretreatment, Extraction and Clean-Up Procedures

While pretreatment of sediments includes drying and sieving, seawater pretreatment usually involves filtering the samples to remove suspended particulate matter. Most of the studies on seawater refer to the soluble fraction; however, some authors analyze the solids retained in the filtration systems [53, 56, 81, 82], or determine the total content, by sonication of seawater before the extraction procedures [75].

Despite the fact that in certain cases seawater samples can be directly injected into the detection system without any type of sample treatment (e.g., surfactants in seawater from China [97] or from the Gulf of Mexico [98]), because pollutant levels in seawater are very low, the extraction techniques applied must be able to concentrate the analytes to improve the sensitivity of the determination.

Table 1 summarizes the extraction techniques used in recent methods recently published for the determination of PAHs, PCBs, OCPs, PPCPs, surfactants and EDCs in seawater. The LODs obtained, the analytical techniques used for detection of compounds and the amounts detected in natural samples are shown.

Table 1. Analytical methods for the determination of PAHs, PCBs, PBDEs, PPCPs, OCPs, EDCs and surfactants in seawater

Analyte	Sample preparation (Sample)	Analytical technique	LODs	Seawater origin	Amounts detected	Ref
Chlorophenols PAHs, PCBs	- Adjust pH (2) and SPE (EN) - SPE (C18)	LC–DAD GC–MS	30 – 200 ng L^{-1} 2 – 3 ng L^{-1}	Gulf of Gdansk (Baltic Sea)	0.1 – 6.0 μg L^{-1} No data	[47]
PAHs	- Filter - SPME (PDMS fiber)	GC–MS	0.1 – 3.5 ng L^{-1}	Jiaozhou Bay (China), 2005	16.5 – 473 ng L^{-1}	[48]
PAHs, PCBs, OCPs	- Filter - SBSE–TD	GC–MS	0.03 – 7.50 ng L^{-1}	Cádiz Bay (Spain)	Phenanthrene, anthracene, pyrene: 100 – 400 ng L^{-1} Others: < 10 ng L^{-1}	[49]
PAHS	- SPE (C18)	LC–UV	0.02 – 0.14 μg L^{-1}	Caspian Sea	0.34 – 14.11ng mL^{-1}	[50]
PAHs, PCBs, NPs, PBDEs	- SBSE–TD - MASE	GC–MS LVI–GC–MS	0.03 – 20.4 ng L^{-1} 0.1 – 222 ng L^{-1}	Getxo (Spain)	PAHs: 9 – 166 ng L^{-1} PCBs: 6 – 44 ng L^{-1} PBDEs: 23 – 214 ng L^{-1}	[51]
PAHs	- Filter - SPE (XAD–4)	GC–MS	0.01 – 2 ng L^{-1}	Deep Bay (South China)	31.7 – 111.8 ng L^{-1}	[52]
PAHs	- Filter - LLE with hexane - Clean-up with silica gel	LC–UV, LC–FD	0.012 ng L^{-1}	Gerlache Inlet sea (Antarctica)	2.1 – 2.8 ng L^{-1}	[53]
PAHs	- LLE with Hex + LLE with DCM - Clean-up silica gel	GC–MS	1.7 – 9.4 ng L^{-1}	Galician coast (Spain), 2002–2004	No data	[54]
PAHS	- Dilute with EtOH (2% v/v) - In–vial MASE	GC–MS	24 – 39 ng L^{-1}	Mallorca (Spain), 2010	Naphthalene: 100 ng L^{-1} Phenanthrene: 91 ng L^{-1}	[55]
PAHs	- Filter - SPE (ENVI–18)	GC–MS	No data	Western Taiwan Strait (WTS), 2009	12.3 – 58.0 ng L^{-1}	[56]
PAHs	- DLLME	GC–MS	1 – 10 ng L^{-1}	Zhanqiao Pier, Shilaoren and Zhanshan beaches (China)	20 to 860 ng L^{-1}	[57]

Table 1. (Continued)

Analyte	Sample preparation (Sample)	Analytical technique	LODs	Seawater origin	Amounts detected	Ref
PAHs	- Filter - Add EtOH	On–line SPE–LC–FLD	0.002 – 0.50 ng L^{-1}	Japan Sea, 2008	Naphthalene: 0.1 – 22.9 ng L^{-1}	[58]
PAHs	- Add IPA - Filter - SPE (C18)	GC–MS	No data	Sozopol harbour, SW Black Sea coast (Bulgaria), 2010	4.9 – 25.1 ng L^{-1}	[59]
PAHs	- Add NaCl - LLE DCM - Clean-up with silica gel	GC–MS	No data	Northern part of the Persian Gulf (Bushehr Province)	1.5 – 3.6 ng L^{-1}	[60]
PAHs	- LLE with DCM	GC–MS	0.02 – 4.2 ng L^{-1}	Coast of Delian (China), 2010	15 – 160 ng L^{-1}	[61]
PAHs	- LLE with Hex	GC–MS	No data	Algeciras Bay (Spain), 2007–2009	10.4 ng L^{-1} – 1.4 μg L^{-1}	[62]
PAHs	- LLE with Hex	LC–UV	0.025 – 0.25 μg L^{-1}	Lebanese marine coast	Mean: 114.5 μg L^{-1}	[63]
PAHs	- Add MeOH	On line SPE–LC–MS/MS		Northern Biscayne Bay, Miami	Naphthalene: 100 μg L^{-1}	[64]
PCBs, OCPs	- SPE (XAD–2) - LLE with Hex - Clean-up with silica gel and florisil	GC–MS	0.02 ng L^{-1}	East coast of India	HCHs: 22.3 – 25.5 ng L^{-1} DDTs: 5.3 – 12.5 ng L^{-1} PCBs: 19.3 – 4.45 ng L^{-1}	[65]
PCBs, PBDEs	- PC–HFME	GC–MS	0.04 – 0.21 ng L^{-1}	Singapore coast	PCBs 0.27 – 4.63 ng L^{-1} PBDEs 1.89 – 21.84 ng L^{-1}	[66]

Analyte	Sample preparation (Sample)	Analytical technique	LODs	Seawater origin	Amounts detected	Ref
PCBs	- Acidify and add $KMnO_4$ - Add H_2O_2 and adjust pH (6) - SPME	GC–MS	0.3 – 7.5 ng L^{-1}	Equatorial Indian Ocean	3.8 – 14.9 ng L^{-1}	[67]
PCBs	- LLE with DCM/Hex - Clean-up with florisil	GC–ECD	1 ng L^{-1}	Northwest of the Persian Gulf, 2010–2011	Cold Season: 0.01 – 0.2 ng mL^{-1} Warm Season: 0.02 – 0.4 ng L^{-1}	[68]
Analgesics, caffeine, TCS, DEET ß–blockers, anti–depressants,	- Filter - Adjust pH (7) - SPE (HLB) - Derivatize (methyl chloromethanoate, only for GC–MS analysis)	GC–MS LC–MS/MS	LOQs: 0.07 – 0.69 ng L^{-1} Semiquantitative	Tromsø–Sound (Norway)	Caffeine: 7 – 87 ng L^{-1} DEET: 0.4 – 13 ng L^{-1} Ibuprofen: < LOQ	[69]
Pharmaceuticals	- Adjust pH (7) - Filter - SPE (HR–X)	LC–MS	LOQs 1 – 50 ng L^{-1}	Belgian coast, 2007–2009	Salicylic: 855 ng L^{-1} Others: 18 – 321 ng L^{-1}	[70]
Pharmaceuticals	- SPE (XAD–2)	LC–MS	1.0 - 13 ng L^{-1}	Central bay of Stockholm	Caffeine: > 25 ng L^{-1}	[71]
Hydroxylated BPs	- DLLME - Derivatize (BSTFA)	GC–MS	32 – 50 ng L^{-1}	Murcia, Benidorm and Alicante (Spain)	BP-3: 1340 – 330 ng L^{-1} BP-1: 280 ng L^{-1}	[72]
UV–filters	- Filter and adjust pH (6) - Add MeOH - SBSE–LD (MeOH)	LC–MS/MS	8 – 120 ng L^{-1}	Ligurian coast (Italy), 2010	BP-3: 88 – 101 ng L^{-1} EHM 25 – 83 ng L^{-1}	[73]
Antibiotics	- Adjust pH (3.5) - SPE (H_2O–Philic disks)	LC–MS/MS	0.2 – 16.7 ng L^{-1}	Polish coastal area, Baltic Sea, 2011–2012	0.6 – 10.8 ng L^{-1}	[74]

Table 1. (Continued)

Analyte	Sample preparation (Sample)	Analytical technique	LODs	Seawater origin	Amounts detected	Ref
UV–filters	- Soluble fraction: Adjust pH (2), filter and DLLME - Total content: Sonicate samples, adjust pH (2) and DLLME	GC–MS	10 – 30 ng L^{-1}	North Adriatic Sea	91 – 750 ng L^{-1}	[75]
Benzophenones	- Adjust pH (2–3) - SPE (HLB) - Derivatize (MSTFA or PFBHA)	GC–MS	0.1 – 1.9 ng L^{-1}	Slovenia and Croatia	2.4 – 380 ng L^{-1}	[76]
OCPs, PCBs	- Filter - SPE (XAD–2) - Clean-up with a florisil	GC–ECD	No data	Gulf of Gdarisk and the Southern Baltic , 1998	Surface: 0.06 – 0.98 ng L^{-1} Botton: 0.09 – 0.95 ng L^{-1}	[77]
OCPs, PCBs	- LLE (Hex) - Clean-up with florisil	GC–ECD	0.2 ng L^{-1}	Mid–Black Sea Coast of Turkey, 1999–2000	PCBs: ND OCPs: 0.2 – 105 ng L^{-1}	[78]
OCPs, PCBs	- LLE (Hex) - Clean-up with silica gel	GC–MS	No data	Singapore, 2003–2004	HCHs: 0.1 – 27.2 ng L^{-1} DDTs: 0.01 – 0.6 ng L^{-1} PCBs: 0.05 – 1.8 ng L^{-1}	[79]
OCPs	- LLE with DCM	GC–ECD	0.5 – 4 ng L^{-1}	Aqaba Gulf (Jordan)	1 pg L^{-1} – 0.392 ng L^{-1}	[80]
OCPs	- Filter - SPE (XAD–4) - Clean-up with silica gel	GC–MS	0.004 – 0.01 ng L^{-1}	Deep Bay (South China), 2004	DDTs: 0.66 – 3.81 ng L^{-1} Average HCHs: 0.71 ng L^{-1}	[81]
OCPs	- Filter - SPE (PAD–3) - Clean-up with silica gel	GC–MS	0.001 – 0.06 pg L^{-1}	From the East China Sea to the high Arctic, 2010	< LOQ – 111 pg L^{-1} (Chlorpyrifos)	[82]
OCPs	- Filter - SPE (XAD–2 and XAD–4, 1:1) - Clean-up with a silica gel	GC–MS	No data	East and South China Seas, 2006	790 (γ-HCH) – 170 pg L^{-1} (CC)	[83]

Analyte	Sample preparation (Sample)	Analytical technique	LODs	Seawater origin	Amounts detected	Ref
OCPs	- Filter - SPE XAD–2 and XAD–4 1:1 - Clean-up with a silica gel	GC–MS/MS	No data	Equatorial Indian Ocean, 2011	0.3 pg mL^{-1} (p,p'-DDE) – 17 ng mL^{-1} (p,p'-DDT)	[84]
OCPs	- LLE DCM - Clean-up with a silica gel	GC–MS	8.5 – 420 pg L^{-1}	Denmark	Only heptachlorepoxide: 3.4-7.5 pg L^{-1}	[85]
BPA, estrogens, APs	- No treatment	On –line SPE HPLC–FD	10 – 30 ng L^{-1}	South Australia		[86]
BPA, APs, pentachlorophenol	- Adjust pH (2) - LLE (DCM) - SPE (HLB) - Derivatize (BSTFA)	GC–MS	3.7 – 15.9 ng L^{-1}	Singapore	BPA: 0.4 μg L^{-1} 4-NP: 20 ng L^{-1} – 2.76 μg L^{-1}	[87]
OCPs, PAHs, Phthalates, adipate	- No treatment	On–line SPE–GC–MS	0.2 – 20 ng L^{-1}	Marina and coast of Salou (Spain)	10 ng L^{-1} (HCB) – 12740 ng L^{-1} (di-*n*-butyl phthalate)	[88]
BPA, APs, chlorophenols	- Acidify - LLE (DCM) - Derivatize (BSTFA) - Clean-up with a florisil	GC–MS	No data	Saemangeum Bay (Korea), 2002–2003.	NP: 298 ng L^{-1}	[89]
fenarimol, fenvalerate, pendimethalin, vinclozolin	- Adjust pH (3) - SPE (SDB disks)	GC–ECD GC–MS (confirmation)	4 – 8 ng L^{-1}	Ionian Sea (Greece), 2003–2004	Vinclozolin: 60 ng L^{-1} Fenarimol: < LOQ	[90]
BPA, APs, chlorophenols,	- Add NaCl - LLE acetone:DCM (20:80, v/v) - Add anhydrous $NaSO_4$ - Derivatize (BSTFA)	GC–MS	0.01 – 0.09 μg L^{-1}	Philippines, 2000–2002	> 1 μg L^{-1}	[91]

Table 1. (Continued)

Analyte	Sample preparation (Sample)	Analytical technique	LODs	Seawater origin	Amounts detected	Ref
APs, APEOs, BPA, estrogens, synthetic steroids	- Filter - SPE (HLB) - Derivatize (BSTFA)	GC–MS	1.9 – 14.1 ng L^{-1}	Thermaikos Gulf (Greece)	NP: 22.0 – 201 ng L^{-1} NPEO: 12.1 – 346 ng L^{-1}	[92, 95]
BPA, APs	- SDME (decanol) - DLLME	LC–UV	4 – 9 µg L^{-1} 0.1 – 1.6 µg L^{-1}	Tenerife (Spain)	< LOQ	[93]
Estrogens, APs, APEOs, BPA	- Filter - Adjust pH (5) - SPE (HLB) - Derivatize (BSTFA)	GC–MS	0.10 – 1.20 ng L^{-1}	Portugal coastal, Atlantic Ocean	Phenols: 8.5 – 1755.9 ng L^{-1} E1, E2, EE2: 1.87 – 8.39 ng L^{-1}	[94]
BPA, APs, estrogens	- Filter - SPE (C18)	LC–FLD	2 – 23 ng L^{-1}	Todos os Santos Bay (Brazil)	4-n-NP: 19.5 – 134 ng L^{-1} E2, E3 y BPA: 4.9 – 76.8 ng L^{-1}	[96]
LAS	- No treatment	FIA–UV	2.4 µg L^{-1}	China	No data	[97]
DOSS, DOSS metabolites, span 80, tween 80, tween 85	- Add IPA (25% , v/v)	LVI–LC–MS/MS	16 – 1250 ng L^{-1}	Gulf of Mexico, 2010	No data	[98]
FAE, AES	- SPE C18 - SPE SAX (separate FAE/AES) - Derivatize (diphenic anhydride)	LC–UV LC–MS	0.5 – 4.8 µg L^{-1}	La Pobla de Farnals beach (Spain)	$C_{12}EO$: 2.3 µg L^{-1} $C_{14}EO$: 99 µg L^{-1}	[99]

Table 2. Analytical methods for the determination of organic pollutants in marine sediments

Analyte	Sample preparation	Analytical technique	LODs	Sediments origin	Amounts detected	Ref
LAS, AES, AS	- Soxhlet: MeOH - PLE: MeOH; 125°C - Clean-up: C18	LC-ESI/MS	1 – 5 ng g^{-1}	Estuary of the river Guadalete, Bay of Cadiz (SW Spain)	168 – 1014 ng g^{-1}	[100]
LAS, AES, AS, NPEOs, AEOs, SPCs, APECs	- PLE: MeOH; 120°C - Clean-up: C18	LC-ESI/MS	1 – 10 ng g^{-1}	Bay of Cadiz (SW Spain)	5370 – 19970 ng g^{-1}	[101]
LAS, AES, NPEOs, AEOs	- PLE: MeOH; 100°C - Clean-up: automated SPE (C18)	LC-MS	1 – 5 ng g^{-1}	Bay of Cadiz (SW Spain)	LAS > 637 ng g^{-1} NPEOs > 401 ng g^{-1} AEO > 861 ng g^{-1} AES >125 ng g^{-1}	[102]
LAS, CDEAs, NPEOs, NPECs, NP, OP	- PLE: MeOH:Ace (1:1) 50°C - Clean-up: C18	LC-ESI/MS	1 – 8 ng g^{-1}	Catalonia (NE Spain)	2.8 – 86940 ng g^{-1}	[103]
QACs	- USE: acidic MeOH - Clean-up: LLE-water + anion exchange resin	LC-ESI-ToF-MS	0.1 – 2.6 ng g^{-1}	Jamaica Bay Bowery Bay	1800 – 74000 ng g^{-1}	[104]
LAS and SOAP	- Soxhlet extraction, MAE, USE. - MeOH (LAS) and petroleum ether (insoluble soap) - Transformation of soap calcium salts into potassium salts - Derivatization with BrMMC and 18-crown-6-ether	LC-FLD	No data	Almeria Coast (Spain)	LAS = 0.04 – 124 µg g^{-1} Insoluble soap = 14 – 3580 µg g^{-1}	[105]
AS and AES	- PLE: MeOH; 125°C	LC-MS/MS	0.04 – 0.09 µg g^{-1}	Almeria Coast (Spain)	1.1 – 29.9 µg g^{-1}	[106]

Table 2. (Continued)

Analyte	Sample preparation	Analytical technique	LODs	Sediments origin	Amounts detected	Ref
AEOs , PEGs	- PLE: MeOH; 120ºC - USE: MeOH:DCM 1:1 - Clean-up: SPE (Oasis HLB) - Automated SPE AutoTrace unit	LC- ESI - MS	< 50 ng g^{-1}	Mar Menor lagoon (Murcia, Spain)	PEGs = up to 9000 ng g^{-1} AEOs < 100 ng g^{-1}	[107]
Fluoroquinolones	- MAME: aqueous HTAB (5%) - Filtration	LC-ESI-MS/MS	0.15 – 0.55 ng g^{-1}	Coastal marine sediments from the southern region of Gran Canaria island (Spain)	3.43 – 206.1 ng g^{-1}	[108]
Pharmaceuticals	- Acidic and parasiticide ivermectin: USE:Ace:acetic acid (20:1) + EtAc (3 times) - Antibiotics: USE:MeOH (2), Ace (1) and EtAc (1) - Clean-up: SPE with Oasis MCX (acidic pharmaceuticals); Lichrolute EN+ Lichrolute C18 (antibiotics); Lichrolute EN (ivermectin)	LC-MS/MS	LOQ = 0.4 – 20 ng g^{-1}	No data	No data	[109]
Analgesic-antiinflamatories, lipid regulators, β-blocker, anti-epileptic drug	- MAME: aqueous POLE 2.75%, 500 W of power - Clean-up: SPE (Oasis HLB)	LC-UV	12 – 556 ng g^{-1}	Gran Canaria Island: Maspalomas (South), the Cicer (NE), Sardina (North) and Taurito (SW)	< LOD	[110]
Atorvastatin, carbamazepine, 17α-ethinylestradiol	- PLE: MeOH at 100ºC	LC-ESI-MS/MS	4.7 – 7.99 ng g^{-1}	Environment Canada Reference site No. 112	No data	[111]

Analyte	Sample preparation	Analytical technique	LODs	Sediments origin	Amounts detected	Ref
Flumequine (FLU), oxolinic acid (OXO)	- MAE: H_3PO4 buffer (1M, pH 2) + DCM - Clean-up: LLE with NaOH (1M).	LC-FLD	FLU = 1.3 ng g^{-1} OXO = 2.4 ng g^{-1}	Catalonia (Spain)	No data	[112]
TCS and two metabolites	- MAE: Ace:MeOH (1:1, v/v); 130ºC - Clean-up: SPE (Oasis HLB)	GC-MS/MS (MTBSTFA)	0.2 ng g^{-1}	NW Spain	< LOD	[113]
Steroid estrogens	- USE: ACN:water (9:1 v/v) - Autoprep EDS-1 cartridge fractionation - GPC: Sephadex LH-20+Florisil and Polyamide columns	LC-ESI-MS/MS	0.02 – 0.32 ng g g^{-1}	Tokyo Bay (Japan)	0.05 – 3.60 ng g^{-1}	[114]
E1, E2, DES	- FMASE: DCM-water	GC-MS/MS	0.0004 – 0.004 ng g^{-1}	Almeria (Spain)	2.5 – 4 ng g^{-1}	[115]
PBs	- SAESC: ACN (3 times)	LC-MS/MS	0.04 – 0.14 ng g^{-1}	Ria Arousa and Ria Pontevedra (Galicia, Spain)	≤ 6.35 ng g^{-1}	[116]
PBs	- SAESC: ACN (3 times) - Clean-up: MISPE	LC-UV LC-MS/MS	1 ng g g^{-1}	- Ria Arousa and Ria Pontevedra - (Galicia, Spain) - Mediterranean Sea (Valencia, Spain)	11.5 ng g^{-1}	[117]
TCS and methyl-TCS	- PLE: DCM; 100ºC - In-cell clean-up: cellulose filters and Na_2SO_4	GC-MS	0.1 ng g^{-1}	Estuary of Guadalete River (SW Spain)	Up to 9.6 ng g^{-1}	[118]
BPs Salicylates	- USE-clean-up (EtAc:MeOH (90:10, v/v), C18)	GC-MS-EI-SIM (BSTFA)	0.07 – 0.28 ng g^{-1}	Mediterranean Coast (Valencia, Spain)	1.4 – 20 ng g^{-1}	[119]

Table 2. (Continued)

Analyte	Sample preparation	Analytical technique	LODs	Sediments origin	Amounts detected	Ref
HBCD diastereoisomers	- MAE: Ace:Hex (1:3); 90ºC - Clean-up: sulfur removal by EPA 3660B standard method, and "acidic washing" with 50% conc. sulphuric acid. - MAE efficiency was compared to Soxhlet and PLE	LC-ESI-ITMS	LOQ = 25 – 40 pg g^{-1}	Marine sediments from oyster culturing sites at Tai-Shi, Bu-dai and Chi-ku (SW coastal Taiwan)	No data	[120]
PCBs	- MAME: Genapol X-080 0.02M and 40% power - Soxhlet: Ace:Hex (10:7 v/v); 50ºC	LC- FLD	0.7 – 3.61 ng g^{-1}	Gran Canaria Island: Las Canteras beach (NE) and Taliarte harbour (SE)	No data	[121]
TBBPA, TCBPA and related phenolic compounds	- MSPD: HCl + Na_2SO_4 + Florisil, transferred to a Florisil column	CE-DAD	LOQ = 182 – 336 ng g^{-1}	NW Spain	No data	[122]
PBDEs	- Soxhlet: Hex:Ace (3:1, v/v); 70ºC - Clean-up: GPC + shaking in H_2SO_4 + silica gel	GC-ECNI-MS	0.1 – 1 ng g^{-1}	Western Scheldt (The Netherlands)	0.1 – 4600 ng g^{-1}	[123]
PBDEs	- PLE–clean-up: Cu + alumina + Hex:DCM (1:1, v/v), 100ºC. - Soxhlet: Hex:DCM (1:1, v/v)	GC-NCI-MS	0.001 – 0.05 ng g^{-1}	-	PBDEs: 0.86-2.34 ng g^{-1}	[124]
PAHs	- USE:DCM - Addition of Copper granules and ashed Na_2SO_4 to remove sulphurous compounds and water, respectively. - Clean-up: columns with 50:50 silica:deactivated alumina.	GC-MS	1 – 2 ng g g^{-1}	Brighton marina (UK)	Total PAHs: 24 – 4710 ng g^{-1}	[125]

Analyte	Sample preparation	Analytical technique	LODs	Sediments origin	Amounts detected	Ref
PAHs	- USE:DCM - Activated copper granules to remove elemental sulphur. - Clean-up: 1:2 alumina/silica gel column and fractionation in hydrocarbons and PAHs	GC-MS	0.2 – 2 ng g g^{-1}	Zhanjiang Bay and Leizhou Bay (South China)	0.1 – 148.2 ng g^{-1}	[126]
OPFRs	- PLE-clean-up: silicon dioxide + Water:ACN (75:25, v/v); 90°C. - Clean-up: SPE (Oasis HLB)	GC-MS	0.2 – 5.1 ng g^{-1}	Marine estuaries in Galicia (NW Spain)	3 – 1900 ng g^{-1}	[127]
OPFRs	- MAE: Ace; 120°C - Clean-up: Oasis HLB + Al_2O_3, EtAc	GC-MS	0.05 – 0.15 ng g^{-1}	Oyster culturing sites at Tai-shi and Chi-ku (SW coastal Taiwan)	0.6 - 4.7 ng g^{-1}	[128]
PCBs	- Soxhlet: Hex - Clean-up: silicic acid/alumina column + silica gel/alumina column	GC-ECNI-MS	10 pg g^{-1}	Sulaibikhat Bay (Kuwait)	1 – 3.056 ng g^{-1}	[129]
Dechlorane compounds DP, Dec-602,- 603,-604	- Soxhlet: Na_2SO_4 + Hex:Ace (1:1, v/v) - Clean-up: H_2SO_4 + silica gel	GC-ECNI-MS	IDL = 0.001 – 0.2 ng g^{-1}	Dalian (NE China)	≤ 1.6 ng g^{-1}	[130]
NBFRs: DBDPE, BTBPE, PBCCH , TBoCT, pTBX, TBPhA, PBT, TDBPP	- PLE: Hex:Ace (3:1, v/v); 70°C - Clean-up: GPC+ Oasis HLB+ silica gel.	GC-ECNI-MS	0.002 – 0.1 ng g^{-1}	Western Scheldt estuary (The Netherlands)	≤ 9.8 ng g^{-1}	[131]
PBDEs and PBBs: Di–decaBDEs	- PLE: DCM:Ace (3:1, v/v); 100°C - Clean-up: LLE in DCM + silica gel + Cu	GC-MS GC-ECD decaBDE	No data	NW head of Tokyo Bay (Japan)	≤ 11.900 µg g^{-1}	[132]

Table 2. (Continued)

Analyte	Sample preparation	Analytical technique	LODs	Sediments origin	Amounts detected	Ref
PBDEs and PBBs: Tetra–hexaBDEs and Tri–decaBDEs	- HD-SPME: water + diluted H_2SO_4 (1:3, v/v) + $KMnO_4$, 100ºC, PA fibre (40 min) - Desorption at 300ºC (2 min)	GC-MS	< 0.15 ng g^{-1}	Marine harbours in Galicia (NW, Spain)	1.4 - 10.3 ng g^{-1}	[133]
PBDEs and PBBs: Mono–octaBDEs	- PLE: DCM; 100ºC - Clean-up: alumina / copper	GC-MS	0.1 – 1.5 ng g^{-1}	San Francisco Estuary (USA)	PBDEs: ≤ 212 ng g^{-1}	[134]
HBCD,TBBPA	- Soxhlet: Ace:Hex (1:1 - 1:3, v/v) - Clean-up: H_2SO_4 + GPC + silica gel	LC-ESI-MS	0.5 – 1.2 ng g^{-1}	Scheldt Basin (Belgium)	0.1 – 9750 ng g^{-1}	[135]
HBCD	- Soxhlet: Ace - Clean-up: acidic aluminium oxide	GC-ECD	0.1 ng g^{-1}	Stockholm (Sweden)	0.1 – 1300 ng g^{-1}	[136]
NBFRs: DBDPE, Deca-BDE	- PLE: DCM; 100ºC - Clean-up: H2SO4 + KOH in EtOH + re-extraction in Hex aminopropyl gel with acid silica.	GC-ECNI-LRMS	0.009 – 0.1 ng g^{-1}	Stockholm archipelago (Sweden)	DBDPE = 0.2 – 11 ng g^{-1} Deca-BDE = 0.55 – 88 ng g^{-1}	[137]
PBDEs	- HS-SPME:water,100ºC, PDMS fibre (30-60 min), desorption (3 min, 280ºC)	GC-IT-MS/MS	0.005 – 0.625 ng g^{-1}	- Catalonia (NE Spain) - Western Scheldt (The Netherlands) - Galicia (NW Spain)	≤ 12 ng g^{-1}	[138]
PBDEs	- USE:MeOH - SBSE: Water dilution, PDMS, 4h). Desorption with ACN, 15 min	GC-MS/MS	0.3 – 203 ng g^{-1}	Northern Portugal	≤ 0.7 ng g^{-1}	[139]
PBDEs PBBs	- MAE:Hex:Ace (1:1, v/v); 152ºC - Clean-up: GPC	GC-IT-MS	0.004 – 0.02 ng g^{-1}	Malvarrosa beach (Valencia, Spain)	BDE-47: 0.2 – 0.3 ng g^{-1}	[140]
DP	- Soxhlet: Cu + DCM - Clean-up: silica gel (multilayer)/alumina	GC-NCI-MS	$4 \cdot 10^{-4}$ – $8 \cdot 10^{-4}$ ng g^{-1}	Yellow sea, Jiaozhou Bay (Qingdao City, China)	0.0004 – 0.126 ng g^{-1}	[141]

Table 3. Analytical methods for the determination of organic pollutants in marine biota

Analyte (number)	Matrix	Sample preparation	Analytical technique	LODs/LOQs	Organism origin	Amounts detected	Ref
17α-ethinylestradiol, estrone	Fish homogenate: - Shorthead redhorse sucker (*Moxostoma macrolepidotum*) for spiked recovery experiments. - Goldfish (*Carassius auratus*)	- PLE: DCM - Clean-up: Envirogel™ GPC columns (lipid removal) - Clean-up: Hex wash (cholesterol removal) - Derivatization: PFBCI	GC-MS (SIM)	EE2 = 0.67 ng g^{-1} E1 = 0.68 ng g^{-1}	Canada	EE2 = 1.53–2.2 ng g^{-1}	[184]
UV filters: 4-MBC, BP-3, EHMC, OC	Fish: - White fish, Coregonus sp. - Roach, *Rutilus rutilus* - Perch, *Perca fluviatilis*	- Homogenization with Na_2SO_4 and column extraction with DCM:Cyclohexane (1:1) - Clean-up: GPC (Biobeads S-X3 column) and silica chromatography	GC-MS	3–380 ng g^{-1} lw	Switzerland	25–166 ng g^{-1} lw	[187]
BPA	Fish: - Bream (*Abramis brama*) - Flounder (*Platichthys flesus*)	- PLE: MeOH, 50°C - Clean-up: NaOH 0.1N + DCM wash and reextraction with DCM	GC-MS/MS	2.6 ng per sample	Wadden Sea (North of The Netherlands)	2–75 ng g^{-1} dw (liver) 1–10 ng g^{-1} dw (muscle)	[183]
4-MBC, OC, Methyl-TCS	Muscle tissue of fish (brown trout, *Salmo trutta fario*)	- Extraction of an aqueous suspension with $K_2C_2O_4$, EtOH, diethylether and *n*-pentane - Clean-up: GPC (Biobeads S-X3 column) and silica chromatography	GC-MS	4-MBC = 5–20 ng g^{-1} lw Methyl-TCS= 2–10 ng g^{-1} lw	Switzerland	50–2400 ng g^{-1} lw	[186]

Table 3. (Continued)

Analyte (number)	Matrix	Sample preparation	Analytical technique	LODs/LOQs	Organism origin	Amounts detected	Ref
17α-ethinylestradiol	Benthic invertebrates: - *Chironomus tentans* - *Hyalella azteca*	- USE: Ace - Filtration through 0.2 µm membrane	LC-MS/MS	No data	Canada	*C. tentans* = 0.5 ± 0.05 µg g^{-1}	[172]
Xenoestrogens: tOP, tBP, BPA	Liver and muscle of rainbow trout *(Oncorhynchus mykiss)*	- MAE: DCM:MeOH, 2:1, 20% of power - Clean-up: SPE (Sep-Pak NH_2)	LC-APCI-MS	10–50 ng g^{-1}	Denmark	19–220 ng g^{-1} wet weight	[176]
Herbicides (29)	- Macroalgae (*Ulva rigida*) - Clams (*Tapes philippinarum*)	- USE: MeOH:water, 4:1 - Clean-up: SPE (Oasis HLB)	LC-ESI-MS/MS	0.012–0.21 ng g^{-1}	Sacca di Goro (Northern Adriatic Sea)	0.13–312.69 ng g^{-1}	[181]
PCBs and chlorinated pesticides	Whale blubber standard reference material (*Globicephala melaena)*	- GC-EDC: Soxhlet extraction (DCM), SEC (preparative-scale divinylbenzene/polystyrene column) - GC-MS: Soxhlet extraction (Hex/Ace), sulphuric acid wash, clean-up: SPE (silica column)	GC-EDC GC-MS	No data	No data	No data	[191]

Analyte (number)	Matrix	Sample preparation	Analytical technique	LODs/LOQs	Organism origin	Amounts detected	Ref
Organo-chlorine compounds	- The gastropod and grazer chink shell (*Lacuna vincta*) - The filter feeding bivalve blue mussel (*Mytilus edulis*) - The predators butterfish (*Pholis gunnellus*) - The seabird black guillemot (*Cepphus grylle*)	- Extraction with Ace:Hex - Clean-up: H_2SO_4 conc	GC-EDC	at least 1 ng g^{-1} for the pesticides and the individual PCB congeners	SW island Flatey in the bay Breiðifjörðu (Iceland)	($ng\ g^{-1}$ lw) Mollusc and fish: ΣPCB = 111–377 ΣDDT = 19–65 HCB = 21–30 Black guillemot (mean) ΣPCB = 2352 ΣDDT = 361 HCB = 283	[168]
PFCs: PFOS, PFOA, PFHxS, PFDS, PFOSA, PFHpA, PFNA, PFDA, PFUnDA, PFDoDA	Liver, kidney, muscle of: - Silver scabbardfish (*Lepidopus caudatus*) and whitemouth croaker (*Micropogonias furnieri*) - Tucuxi dolphins (*Sotalia guianensis*) - Brown mussel (*Perna perna*) - Tilapia (*Tilapia rendalli*), pearl cichlid (*Geophagus brasiliensis*) - Mullet	- Liver and kidney: Ion-pair extraction - Muscle: Alkali-digestion (0.01N KOH/MeOH) - Clean-up: SPE (Oasis HLB or Oasis WAX)	LC-ESI-MS/MS	No data	SE Brazil	0.47–109.3 $ng\ g^{-1}$ wet weight	[196]

Table 3. (Continued)

Analyte (number)	Matrix	Sample preparation	Analytical technique	LODs/LOQs	Organism origin	Amounts detected	Ref
PFCs: PFOS, PFHS, PFBS, PFOSA, PFDoDA, PFUnDA, PFDA, PFNA, PFOA, PFHpA, PFHxA	- Green-lipped mussels, *Perna viridis* - Oysters, *Crassostrea gigas*	- Alkali digestion (0.01N KOH/MeOH) - Clean-up: SPE (Oasis HLB), two fractions	CG-MS/MS	No data	South China and Japan	103.2–3,996.9 pg g^{-1} dw	[197]
PAHs (9), PCBs (7), DDT, DDE, DDD	- Freeze-dried mussel tissue standard reference material - SRM2974 (NIST). Mussel tissue homogenate recommended control material IAEA-142 (IAEA) - Freeze-dried Antarctic krill samples (ISS, Rome, Italy).	- Digestion with KOH (20% MeOH) in ultrasonic bath, 60°C - USE (Hex) - Clean-up system, elution with Hex and supercritical CO_2	GC-MS-SIM	0.03–0.2 ng mL^{-1}	No data	0.03–170 ng g^{-1} dw	[221]
Paroxetine, fluoxetine, norfluoxetine	Fish homogenate: - Brownbullhead (*Ameiurus nebulosus*) - Gizzard shad (*Dorosoma cepedianum*) - White perch (*Morone americana*)	- PLE (MeOH,100°C) - Clean-up: SPE (Oasis MCX)	LC-APCI-MS/MS	Paroxetine = 0.07 ng g^{-1} Fluoxetine = 0.02 ng g^{-1} Norfluoxetine = 0.04 ng g^{-1}	Canada	Paroxetine = 0.48–0.58 ng g^{-1} Fluoxetine = 0.20–1.02 ng g^{-1} Norfluoxetine = 0.15–1-08 ng g^{-1} (wet weigth)	[182]

Analyte (number)	Matrix	Sample preparation	Analytical technique	LODs/LOQs	Organism origin	Amounts detected	Ref
Antidepressants: fluoxetine, norfluoxetine, sertraline, desmethyl-sertraline	Liver, brain, and lateral filet tissues of fish: - *Lepomis macrochirus* (blue gill) - *Ictalurus punctatus* (channel catfish) - *Cyprinus carpio* - *Pomoxis nigromaculatus* (black crappie)	- Dilution of homogenates (1:5 with 0.1M phosphate buffer, pH 6) - Vortex agitation with ice-cold ACN - Clean-up: column SPE (Bond Elute Certify) - Derivatization: pentafluoropropionic anhydride	GC-MS	0.01 ng g^{-1}	Denton County, North Texas (USA)	> 0.1 ng g^{-1}	[215]
Antidepressant pharmaceuticals (10)	Fish neural tissue: brain tissue from native white suckers (*Catostomus commersoni*)	- Brain homogenization (+ formic acid and ACN) with ultrasonic tissue disruptor - Centrifugation and supernatant analysis	LC-MS/MS	0.015 ng g^{-1} (LOQ)	Colorado and Iowa (USA)	0.02–3 ng g^{-1}	[170]
Barbital homologous residues (7)	Fish muscle	- USE (0.1% formic acid in ACN) - Clean-up: *n*-hexane wash, SPE (Oasis HLB)	LC-MS/MS	0.25–0.50 ng g^{-1} (LOQ)	Changsha (Hunan, China)	0.3–1.5 ng g^{-1}	[174]

Table 3. (Continued)

Analyte (number)	Matrix	Sample preparation	Analytical technique	LODs/LOQs	Organism source	Amounts detected	Ref
Carbamazepine, clorfibric acid, ketoprofen, naproxen, bezafibrate and ibuprofen	- Mussel samples (*Mytilus galloprovincialis*): for method optimization. - Clams (*Ruditapes philippinarum*) - Donax clams (*Scrobicularia plana*) - Oysters (*Spondylus calcifer*)	- MAME (POLE 3%, 500 W power) - Clean-up: SPE (Oasis HLB) - Comparison with Soxhlet extraction (MeOH) for validation purposes	LC-DAD	< 0.22 μg g^{-1}	Gran Canaria Island (Spain)	No data	[177]
Diclofenac, ibuprofen, propanolol	Baltic Sea blue mussels, *Mytilus edulis trossulus*	- Vortex agitation: 5% ammonia in 80% MeOH (diclofenac) and 2% formic acid in MeOH (propranolol) - Clean-up: *n*-hexane wash, SPE (Oasis MAX)	LC-MS/MS	No data	Askö, Baltic Sea (Austria)	Diclofenac = 0.16 μg g^{-1} Propanolol = 0.18 μg g^{-1}	[165]
Pharmaceuticals: Diclofenac, naproxen, ibuprofen, bisoprolol, and carbamazepine	Blood plasma and bile from rainbow trout (*Oncorhynchus mykiss*)	- SPE (Oasis HLB)	LC-MS/MS	Plasma = 0.86–20 ng mL^{-1} Bile = 20–100 ng mL^{-1}	Savon Taimen (Finland)	Plasma = 0.62–210 ng mL^{-1}	[159]
Diclofenac, mefenamic acid, trimethoprim, carbamazepine gemfibrozil	Marine blue mussels: - *Mytilus edulis* - *Mytilus galloprovincialis*	- PLE (ACN: water, 3:1; 60°C) - Clean-up: SPE (Strata-X)	LC-MS/MS	4–29 ng g^{-1} (LOQ)	Ireland coast	< 23 ng g^{-1}	[180]

Analyte (number)	Matrix	Sample preparation	Analytical technique	LODs /LOQs	Organism source	Amounts detected	Ref
Diltiazem, diphenhydramine, carbamazepine, fluoxetine, norfluoxetine	Muscle of white sucker (*Catostomus commersoni*) and johnny darter (*Etheostoma nigrum*)	- *In-vivo* and *in-vitro* SPME (PDMS fiber)	LC-MS/MS	0.2–50.4 pg g^{-1}	Ontario (Canada)	2.04–81.63 pg g^{-1}	[195]
Pharmaceuticals and metabolites	Fish tissue (*Lepomis* sp.)	- Extraction by agitation (acetic acid, pH 4:MeOH)	LC-MS/MS	0.01–3.14 ng g^{-1}	Denton County, Texas (USA)	0.11–5.14 ng g^{-1}	[158]
Pharmaceuticals (24) PCPs (12)	Fish fillet (including skin and belly flap) and liver tissues	- USE: 0.1M acetic acid:MeOH, 1:1, v/v Extraction with Ace Clean-up with GPC (silica gel) and derivatization.	LC-MS/MS GC-MS/MS	0.01–3.14 ng g^{-1} (MDL) 5.1–397 ng g^{-1}	USA	Up to 2100 ng g^{-1}	[166]
Pharmaceuticals (6) and personal care products (12)	Tilapia fillets, small mouth bass fillets, and bream fillet	- PLE: DCM:EtAc, 1:1, v/v; 80°C + silica gel clean-up - Clean-up: GPC	GC- MS/MS	1.2–38 ng g^{-1} (MDL)	USA and German Environmental Specimen Bank	5.5–81.2 ng g^{-1}	[185]
Pharmaceuticals (13)	Unicellular algae *Pseudokirchneriella subcapitata*	- USE: MeOH, 2% ammonia in MeOH, Ace - Clean-up: SPE (Oasis HLB and Oasis MCX)	HPLC-MS/MS	low ng L^{-1} range	Collection of Algal Cultures, Göttingen, Germany	No data	[173]
Pharmaceuticals: Atorvastatin, carbamazepine, EE2, triclosan	Benthic invertebrates, the midge Chironomus tentans and the amphipod Hyalella azteca	- USE: Ace	LC-MS/MS	0.3–7.7 ng mL^{-1}	Canada	No data	[223]

Table 3. (Conitnued)

Analyte (number)	Matrix	Sample preparation	Analytical technique	LODs /LOQs	Organism source	Amounts detected	Ref
Pharmaceuticals: Atorvastatin, carbamazepine, EE2, triclosan	Benthic invertebrates, the midge *Chironomus tentans* and the amphipod *Hyalella azteca*	- USE: Ace	LC-MS/MS	0.3–7.7 ng mL^{-1}	Canada	No data	[223]
Emerging contaminants (10)	Fish muscle (dorsal–epaxial muscle): - Rainbow Trout (*Oncorhynchus mykiss*) - Greenside Darter (*Etheostoma blennioides*)	- SLE: 95:5:0.1 (v/v/v) MeOH:water:formic acid - *In-vivo* analysis: SPME (PDMS)	LC-MS/MS	Fish SPME = 0.5–15 ng g^{-1} (MDL)	Ontario (Canada)	0.53–175 ng g^{-1}	[194]
Diclofenac	Bile of Rainbow Trout (*Oncorhynchus mykiss*)	- SPE: Strata-X-AW	UPLC-ToF-MS	No data	Dorset, UK	328–12,724 ng mL^{-1}	[161]
Diclofenac and metabolites	Fish bile of Rainbow trout (*Oncorhynchus mykiss)*	- Dilution and SPE (Oasis HLB)	LC-UV LC-IT, LC-Q-ToF (metabolites identification) LC-MS/MS (quantification)	No data	Rautalampi (Finland)	No data	[160]

Analyte (number)	Matrix	Sample preparation (Sample)	Analytical technique	LODs/ LOQs	Organism source	Amounts detected	Ref
Ibuprofen	Blood plasma and tissues (muscle, gill, liver) of fish: - Fathead minnow (*Pimephales promelas*) - Channel catfish (*Ictalurus punctatus*)	- Blood plasma: LLE (Ace) - Tissue: Extraction of Blot-dried tissue in a Mini-Beadbeater™. Clean-up: SPE (ResPrep™ florisil) - Derivatization: BF3/MeOH	Ibuprofen: GC-MS Metabolites: LC-MS	No data	University of North Texas, Aquatic Toxicology facility and Pond King Inc., Gainesville, Texas (USA)	Fathead minnow = 45.99–173 ng g^{-1} wet weight Catfish = 18.74–314.8 ng g^{-1} wet weight	[164]
Propanolol	Blood plasma of rainbow trout *(Oncorhynchus mykiss)*	- Plasma: vortex agitation with NaOH 0.1% + EtAc	LC-MS	0.025 ng mL^{-1}	No data	0.94–5200 µg mL^{-1}	[163]
Carbamazepine	- Crustacean *Thamnocephalus platyurus* - Algae *Pseudokirchneriella subcapitata*	No data	LC-ESI-MS/MS	0.2 ng L^{-1}	Canada	129 ± 57 µg mg^{-1}	[218]
Veterinary quinolone antibiotics (5)	Fish muscle: - Gilthead (Greece) - Salmon (Norway) - Trout (Rioja, Spain) - Sea bass (Santa Pola, Spain) - Mussel (Galicia, Spain) - Prawn (Honduras), - Turbot (Galicia)	- Homogenized sample + 0.05M SDS, pH 3 - Extraction by agitation - Filtration of supernatant (0.45 µm, nylon membranes)	Micellar LC-FLD (LAS)	1–7 ng g^{-1}	No data	Not detected in any of the analyzed samples	[207]

Table 3. (Continued)

Analyte (number)	Matrix	Sample preparation (Sample)	Analytical technique	LODs/ LOQs	Organism source	Amounts detected	Ref
Flumequine, Nalidixic acid, Oxolonic acid	Schrimps: - *Pandalus spp.* - *Penaeus vannamei* - *Penaeus monodon* - *Pandalus borealis*	- Homogenized at high speed (ACN) - Degrasing: vortex agitation of extract + 0.1 M ammonia + Hex - LLE: 0.1M HCl + EtAc	HPLC-FLD	Nalidixic acid = 6.9 ng g^{-1}	Northwest and Northeast Atlantic	88.5–100 ng g^{-1}	[208]
Quinolone antibiotic: flumequine	Muscle tissue of fish	- Samples were stirred (disintegrated) with buffer pH 9.1 - USE: ACN - Defating with Hex	HPLC-FLD	1 ng g^{-1}	Slovenia (Poland)	No data	[169]
Antibiotics: oxytetracycline (OTC), flumequine (FLU), oxolinic acid (OA)	Bryophyte *Fontinalis antipyretica* Hedw.	- OA, FLU: sample + oxalic acid in EtAC and soaked in liquid nitrogen. Centrifugation and extraction with TCM - OTC: sample soaked in liquid nitrogen. Extraction with acetone + McIlvaine buffer + EDTA (sat). Clean-up: SPE (BondElut C18)	HPLC-UV	5–30 ng g^{-1}	Chézine Valley (Nantes, Loire-Atlantique, West of France)	No data	[206]
Aminoglycoside residues (13)	Fish	- Homogenization with TCA solution + EDTA - Clean-up: SPE (Oasis MCX) aconditiioned with LiOH	UPLC-MS/MS	2–25 ng g^{-1}	Switzerland	No data	[219]

Analyte (number)	Matrix	Sample preparation (Sample)	Analytical technique	LODs/ LOQs	Organism source	Amounts detected	Ref
Antibiotics (11) and five main metabolites	Fish tissue and mussel samples: Hake (*Merluccius merluccius*) Anchovy (*Engraulis encrasicolus*) Mussel (*Mytilus* sp.) Wedge sole (*Solea solea*)	- MAE with Proteinase-K (5 min, 50 W of power). - Extract agitation with formic acid + DCM	HPLC-MS	2–16 ng g^{-1}	Seville (Spain)	Carp = 20–970 ng g^{-1} Wild fish = 10–590 ng g^{-1} (viscera and tissue)	[178]

Liquid-liquid extraction and solid phase extraction have been the most widely used. In some cases, to reduce solvent amounts and to increase concentration factors, microextraction techniques such as solid-phase microextraction, stir-bar sorptive extraction, dispersive liquid-liquid microextraction, polymer-coated hollow fiber microextraction, single drop microextraction and membrane assisted solvent extraction and have also been proposed.

Liquid-liquid extraction. It is well known that LLE is a time-consuming technique that often requires large volumes of organic solvents and is difficult to automate. Despite this, LLE has been widely used and was used with Hex for the determination of several PAHs in seawater from Gerlache Inlet Sea [53], the Algeciras Bay (Cádiz) [62], and the Lebanese marine coast [63]. In the first case, data confirmed the still pristine Antarctic environment. In the Algeciras Bay, high concentrations of PAHs were found and in samples from the Lebanese marine coast, among the three analyzed PAHs (phenanthrene, anthracene and pyrene), only pyrene was quantified. LLE with DCM was used for the determination of 14 PAHs in seawater from Northern part of the Persian Gulf (Bushehr Province) [60] and 24 PAHs in seawater samples taken 50 days after the oil spill along the Dalian coast (China) [61]. In addition, LLE in combination with Hex/DCM was used for the determination of 30 PAHs in seawater from the Galician coasts after the Prestige spill [54]. No data about concentrations are provided and the study focused on assessing the major sources of PAHs into seawater. LLE has been also used to extract OCPs and PCBs. Thus, LLE with DCM was used for the determination of 3 families of OCPs (DDT and its metabolites, HCC isomers, aldrin and heptaclor) in addition to HCB, in seawater from the Gulf of Agaba, Jordan [80]. *p,p'*-DDT and β-HCH were detected in all samples. DCM was also employed for the determination of 3 OCPs (cypermethrin, heptchlor and heptachlor epoxide) in seawater from Denmark [85]. Neither cypermethrin nor heptachlor were found in the analyzed samples while heptachlor epoxide had a detection frequency of 100%. A mixture DCM/Hex was used for the determination of six PCBs in seawater from the Norhwest of the Persian Gulf, sampled in two different seasons [68]. In addition, Hex was used for the determination of several OCPs and PCBs in seawater along the mid-Black Sea Coast of Turkey [78] and seawater from Singapore [79]. PCBs were not detected in any of the Turkish samples, whereas OCPs was detected at high concentrations. In Singapore, HCHs, DDTs and PCBs were quantified. For EDCs, LLE with DCM have been successfully used to determine BPA, APs, and chlorophenols in seawater from Singapore coastline [87] and from the Saemangeum Bay in Korea [89].

In Singapore, the most commonly AP was 4-NP, whereas BPA was detected in the majority of the samples. In samples from Korea, most analytes were not detected except NP that was found at ng L^{-1} levels. In addition, a mixture DCM:Ace was employed to determine the same compounds in seawater from different Philippines Bays [91]. Among the 11 studied phenols, 4-*n*-NP, pentachlorophenol, BPA, 4-*t*-BP, 4-*t*-OP and 2,4-dichlorophenol were detected in most of the analyzed samples.

Solid-phase extraction. This technique has been widely used for the analysis of seawater. Although SPE can be used off-line, on continuous, or coupled on-line to a chromatographic technique, the off-line mode was the one chosen for most of the studies. SPE with octadecyl silica sorbents (C18) has been particularly used for the analysis of PAHs. C18 sorbent has been used to study PAHs and PCBs in Coastal waters from the Gulf of Gdansk, Baltic Sea [47]. Phenanthrene, benzo(b)fluoranthene and anthracene were found at ng L^{-1} levels. Chlorophenols were quantified at low μg L^{-1}, whereas concentrations of PCBs did not exceed few ng L^{-1}. C18 sorbents were also employed to determine seven PAHs in seawater from the Caspian Sea [50]. Naphtalene, acenaphtene, fluorene, phenanthrene, anthracene, fluoranthene and pyrene were found. C18 was also the sorbent used to determine 16 PAHs in seawater from Sozopol Harbor, Bulgaria. Naphthalene, anthracene, phenanthrene, acenaphthylene, acenaphthene, fluorine, fluoranthene and pyrene were quantified [59]. This sorbent was also used for the determination of eight EDCs in samples collected from Todos os Santos Bay, Brazil [96], where BPA, 4-*n*-NP, E2 and E3 were detected; and surfactants, FAEs and AEOs, from seawater collected at a beach near the urban area of La Pobla de Farnals [99].

XAD-2 and XAD-4 (polystyrene-divinylbenzene) were the most widely polymeric sorbents used for the analysis of OCPs. They were also applied to the analysis of PAHs, PCBs and pharmaceuticals. For example, XAD-2 resin was used to determine DDT and its metabolites, HCH isomers, HCB and seven PCBs in the Gulf of Gdarisk and the Southern Baltic [77]; and nine PCBs, DDT, DDT metabolites and HCH isomers from seawater in the East Coast of India [65]. In addition, SPE performed with XAD-2 bags instead of columns was employed to analyze ten pharmaceuticals in seawater from Central Bay of Stockholm [71]. Among them, caffeine, metoprolol, oxazepam and carbamazepine were detected, with caffeine being the compound found at highest concentrations (> 25 ng/L). SPE using XAD-4 resin was successfully applied to determine several OCPs [81] and PAHs [52] in Deep Bay, South China. DDT and metabolites were detected in all samples. A mixture of XAD-

2 and XAD-4 (1:1) was also employed to analyze several OCPs including DDT, DDE, HCH isomers, cis-chlordane (CC) and trans-chlordane (TC) in East and South China Seas [83] and in the Equatorial Indian Ocean [84]. In seawater from China, DDTs, HCHs and chlordane were detected in all samples and in water from the Equatorial Indian Ocean, *pp'*-DDT was the predominant compound. HLB has been selected as the optimal SPE sorbent to study EDCs in seawater. Thus, it has been employed to determine BPA, APs, APEOs, and several estrogens and synthetic steroids in samples from the Thermaikos Gulf, Greece [92, 95]. Most of the phenolic compounds (*t*-OP, OP_2EO, NP, NP_1EO, NP_2EO and BPA) were identified, whereas steroid EDCs were not detected, with NP and NPEO being the dominant compounds. HLB was also used to extract BPA, APs and their metabolites as well as several estrogens from seawater from Portugal coastal areas (Atlantic Ocean) [94] where all the analyzed compound were detected; and for the analysis of PPCPs (analgesics, β-blockers and anti-depressants, caffeine, TCS) and the insecticide DEET in seawater from Tromsø-Sound, Norway [69]. Among the studied compounds, DEET, caffeine, ibuprofen together with its metabolites were detected. In addition, HLB was also selected for the determination of 8 BPs in seawater from Slovenia and Croatia [76]; being BP-3 the most frequently detected compound.

Less commonly used sorbents used to analyze pollutants in seawater include EN, ENVI-18, HR-X, SDB-2 and PAD-3. SPE using EN sorbent was used to extract chlorophenols in coastal waters of the Gulf of Gdansk, Baltic Sea [47]. The bonded octadecyl ENVI-18 sorbent was used to extract 15 PAHs from the Wester Taiwan Strait [56]. The hydrophobic polystyrene-divinylbenzene sorbent HR-X was successfully used to determine 13 pharmaceuticals in Belgian coastal areas [70]. Salicylic acid was detected in more than 90% of the samples at concentrations up to 855 ng L^{-1}. Finally, PAD-3 (macroporous copolymerisates of polystyrene and divynilbenzene) was employed to extract six OCPs from seawater collected from the East China Sea to the High Arctic [82]. The insecticides α-endosulfan, chlorpyrifos and dicofol were the most abundant chemical.

SPE can also be performed using hydrophilic speed disks which were used for the analysis of eleven antibiotics in seawater from the Polish coastal area of the Southern Baltic Sea [74]. Among the analyzed antibiotics, sulfathiazole, sulfapyridine, enrofloxacin, sulfamethazine and sulfamethiazole were detected but not quantified, and trimethoprim, sulfamethoxazole, sulfadimethoxine were quantified. Styrene-divinylbenzene SDB disks were used to extract four pesticides considered as EDCs from samples from the Ionian Sea, Greece [90]. Only fenarimol and vinclozolin were found in the analyzed samples.

Because of its simplicity, on-line SPE coupled to LC is a good choice to analyze pollutants in seawater. Since the evaporation and reconstitution steps, usually present in off-line SPE are eliminated, on-line methods are less time-consuming than off-line ones. This technique was successfully applied to analyze 15 PAHs in seawater from Japan [58], 16 PAHs and alkylated PAHs from Northern Biscayne Bay, Miami [64] and five EDCs in sediments and seawater from South Australia [86]. In Japanese seawater, naphtalene was the predominant compound. In Miami, only napthalene was found, while in South Australia, a degradation study found that the five studied ECDS degraded rapidly under aerobic conditions, being more persistent under anoxic conditions. On-line SPE can also be used with GC. This technique was used to determine 14 EDCs in marine and coastal waters from Salou (Spain) [88]. Phthalates [di-*n*-butyl phthalate, benzyl butyl phthalate and bis (2-ethylhexyl) phthalate], HCB and bis-[2-ethylhexyl] adipate were found.

Microextraction techniques. SPME is based on the partition equilibrium of the analyte between the sample and the sorbent. In combination with GC, SPME is particularly suitable for the determination of volatile and semi-volatile non-polar compounds. Although SPME minimizes the disadvantages of conventional techniques, it has seldom been used for the determination of pollutants in seawater. SPME was used to determine 16 PAHs in water from the Jiazhou Bay, China [48] and eight PCBs in water from the Equatorial Indian Ocean [67]. In the first case, naphtalene, fluorene, anthracene and fluoranthene were found, whereas in the Equatorial Indian Ocean, PCBs 1, 5, 47, 98, 154, 171 and 201 were quantified.

SBSE is based on the same principles of SPME with the advantage of a much larger polymer coating of polydimethylsiloxane that result in higher sample capacity and extraction efficiency. Once the extraction concludes, analytes must be thermally or chemically desorbed with a solvent. Thermal desorption was successfully applied to the determination of PAHs, PCBs and OCPs in water samples from Cadiz Bay, Spain [49] and PAHs, PCBs, NPs and PBDEs in seawater from Getxo, Spain [51]; while chemical desorption was used for the determination of six UV filters in seawater along the Ligurian coast of Italy [73]. In Cadiz Bay, phenanthrene, anthracene and pyrene were detected at all sampling points. Other PAHs and DDT were also detected in several samples taken close to wastewater discharge points. In Getxo, high amounts of PAHs, PCBs and PBDEs were quantified, while NPs were not found in any of the analyzed samples. In the Ligurian coast, only BP-3 and EHMC were detected in the analyzed samples.

DLLME is an extraction technique developed in recent years. The extraction of the analytes takes place in a dispersion of an extracting solvent in water. To achieve the dispersion, a second solvent (dispersing solvent) is used. This technique was applied to the determination of 16 PAHs in water from different Chinese beaches [57], four BPs in water from different beaches in Murcia, Benidorm and Alicante (Spain) [72], eight UV filters in water from the North Adriatic Sea [75], and five EDCs in the coasts of Tenerife (Spain) and in the vicinity of a petroleum refinery [93]. In seawater from Chinese beaches, all studied PAHs were found; whereas in Spanish beaches, BP-3 was found in all analyzed samples, BP-1 in one of the analyzed samples, and BP-8 and 2,3,4-trihydroxy-benzophenone were not found. In the North Adriatic Sea, particulate and soluble fractions were analyzed and all target compounds were detected. Finally, in Tenerife, only 4-t-BP and 4-n-NP were detected, but not quantified.

SDME involves the extraction of the analytes with a single microdrop of solvent suspended at the tip of a syringe. Together with the DLLME method, a SDME method was also developed for the determination of EDCs in the coasts of Tenerife [93], obtaining LODs 20 times higher than with DLLME alone.

PC-HFME combines SPME and SBSE on a hollow fiber. This novel technique was successfully applied for the determination of six PCBs and two PBDEs from coastal seawater in Singapore [66].

In MASE, the extraction occurs by diffusive transport through the pores of an organic solvent-impregnated membrane. Two different methods, MASE combined with large volume injection and SBSE, were compared and applied to the determination of PAHs, PCBs, NPs and PBDEs in seawater from Getxo, Spain [51]. LODs using MASE were up to ten times higher than those using SBSE. Regarding the application to seawater samples, comparable results were obtained for both procedures. In addition, a variation of the procedure, in vial-MASE, was used to extract seven PAHs from seawater collected in Palma de Mallorca's Harbour and Pollensa's Harbour (Mallorca Island, Spain) [55]. Only one of the analyzed samples contained naphthalene and phenanthrene at concentration higher than the LOQ.

Regarding marine sediments, since the analyte-matrix interactions are much stronger than those usually occurring in seawaters, the techniques traditionally used for isolation of the contaminants have been more robust. Soxhlet, PLE, USE, MAE, MAME, FMASE, SAESC or MSPD, HS–SPME and HD–SPME have been widely used. Given the ability to extract matrix components of these techniques, and to minimize matrix effects, after extraction, SPE, LLE and GPC have traditionally been used as sample clean-up techniques. Table 2 summarizes the extraction techniques used in methods recently published in the literature for the determination of pollutants in marine sediments. The LODs obtained, the analytical techniques used for detection of compounds, and the amounts detected in natural samples are shown.

Soxhlet has been used for de determination of surfactants (LAS, AES and AS) in the estuary of the river Guadalete in the Bay of Cadiz [100] and in Almeria Coast (LAS and insoluble soap) [105] and all of the analyzed homologues were quantified. PCBs were analyzed in the Canary Islands [121] and in the Sulaibikhat Bay of Kuwait. [129]. No data of contamination were reported in the first location, and in Kuwait all the studied compounds were quantified in at least one sample except for the PCB congeners 169 and 201. Tetra–decaBDEs were analyzed in the Western Scheldt, The Netherlands [123], and 39 PBDEs [124] were also analyzed using Soxhlet, five of them were found in at least one sample (BDE-47, 100, 99, 118, 154, 153 and 183). Additionally, ten dechlorane derivatives, DP, Dec-602, -603, -604 [130] were analyzed in Dalian, North-East China and HBCD and TBBPA were analyzed in the Scheldt Basin in the coast of Belgium, HBCD in Stockholm [135, 136 where all compounds were detected and quantified. DP was also determined in Jiaozhou Bay, China [141].

PLE was also used for the determination of surfactants in different Spanish locations including the Bay of Cadiz, [100-102], Catalonia [103], Almeria Coast (Spain) [106] and Mar Menor lagoon in Murcia [107]. In all cases, high amounts of surfactants were quantified. Some pharmaceuticals (atorvastatin, carbamazepine and 17α-ethinylestradiol) have been studied in marine enviroments of Canada [111]. No data of contamination are reported in the referenced work.

TCS and methyl-TCS were determined in the Estuary of the Guadalete River, in the South-West of Spain [118]. All samples were contaminated with both compounds. Seven organophosphate flame retardants were studied in different marine estuaries of Galicia, in the North-West of Spain [127]. Several sediments were subjected to the optimized procedure and in most of the samples the compounds remained below the LOQs of the method; however, relatively low levels of pollutants (TiBP, TCEP; TPP and TCPP) were detected in two of the analyzed samples. NBFRs were determined in the Western Scheldt estuary in the Netherlands [131]. All the compounds, with the exception of TBPhA, TDBPP, BDE 3, BDE 156, were detected in samples. Two NBFRs (DBDPE, Deca-BDE) were also studied in the Stockholm archipelago [137]. Both compounds were quantified in high concentrations. Finally, PLE was applied to the determination of PBDEs and PBBs in marine sediment collected from the Tokio Bay [132] and in the San Francisco Estuary [134] where almost all compounds tested were detected.

USE has been widely used for the determination of pollutants in marine sediments including the determination of 12 QACs in the Bowery Bay in Jamaica [104] and different surfactants in Spanish coasts, LAS and insoluble soap in Almería [105], and AEOs and PEGs in the Mar Menor lagoon in Murcia [107]. Acidic pharmaceuticals, parasiticide ivermectin and some antibiotics [109] were also studied using USE, but no data on contamination were reported. Twelve steroids were analyzed in 20 samples of marine sediments [114]. E1 was found in all samples, β-E2 in a high percentage of them, α-E2 in seven and E1-3S in only three of the samples. Otherwise, UV-filters (five BPs and two salicylates) were analyzed in the coast of Valencia, Spain [119]. Only EHS and DHDMB were detected and quantified. PAHs were analyzed in the Brighton Marina, UK [125] and in Zhanjiang Bay and Leizhou Bay, South China [126] where all the studied PAHs were detected and quantified. Finally, USE in combination with SBSE was used for the determination of nine PBDEs in the Northern Coast of Portugal [139]. No data on the use of the optimized method to marine sediment samples were reported.

MAE has also been used for the determination of surfactants (LAS and insoluble soap) [105]. Quinolones [112], TCS and two metabolites (2,4-DCP and 2,4,6-TCP) [113] were analyzed in different locations of the coast of Catalonia and Galicia in Spain. While in the former location no data on contamination were reported, in the latter location, although significant concentrations were found in several environmental compartments such as sewage sludge or river sediments, no contamination was detected in marine sediments. HBCD diastereoisomers were determined in marine sediments from oyster culture sites in Tai-Shi, Bu-dai and Chi-ku in South-West Taiwan [120]. This study compared the efficiency of MAE, Soxhlet and PLE. However, no data on contamination are reported in the referenced paper. This study also analyzed some OPFRs [128]. Five of the studied compounds were found in samples (TCEP, TCPP, MDPP, TDCPP and TPP). PBDEs were analyzed in the Malvarrosa beach (Valencia, Spain) [140] and high concentrations of BDE 47 were found. The variant of the MAE, called MAME, was used for the determination of five fluoroquinolones [108], eight pharmaceuticals, including analgesic-antiinflamatories, lipid regulators, β-blockers, anti-epileptic drugs [110], and seven PCBs in marine sediments from the Coast of Gran Canaria, Spain [121]. Variable amounts of the five studied fluoroquinolones (levofloxacin, norfoloxacin, ciprofloxacin, enrofloxacin and sarafloxacin) were found, but all pharmaceuticals were detected at concentrations below the LOQ. No data of contamination with PCBs were reported.

Other techniques such as FMASE for the determination of estrogens [115], SAESC for the determination of PBs [116, 117], MSPD for the determination of TBBPA, TCBPA and related phenolic compounds [122], HS–SPME for the determination of PBDEs and PBBs [133], and HD–SPME for PBDEs [138] have also been used but to a lesser extent.

For clean-up, SPE, both automated on-line or manual modes, have been widely applied. C18 cartridges [100-103, 109], Oasis HLB [107, 110, 113, 127, 128], Oasis MCX [109] and Lichrolute EN with Lichrolute EN [109] has been used. LLE [104, 112, 132] and other techniques such as GPC [114, 123, 131, 135, 140], filtration [108], MISPE [117], MSPD [122] or clean-up with silica/alumina mixtures [125, 126, 129, 141], silica gel [130], alumina [134, 136] or C18 with solvents [119], have seldom been used.

3.2. Analytical Techniques

Despite performing the sample treatment, an appropriate analytical technique must be selected to analyze the target compounds. Table 1 and Table 2 also show the analytical techniques most commonly used for the detection and quantification of pollutants in seawater and marine sediments.

GC coupled to MS has been, by far, the most commonly used technique. PAHs, PCBs, PBDEs have been analyzed by GC-MS [47-49, 51, 52, 54-56, 59-62, 65-68, 72, 75, 76, 79, 81-85, 87-92, 94, 95]. For other pollutants such as BPs, BPA, APs, estrogens and synthetic steroids, a derivatization step prior to GC analysis is required [69, 72, 76 89, 91, 92, 94, 95], this technique has been also widely used in sediments [113, 115, 118, 119, 123-134, 137-141]. Since GC-ECD responds adequately to molecules with electronegative atoms like halogens, it has been used to determine OCPs and PCBs [68, 77, 78, 80, 90, 132, 136].

Since the majority of PPCPs and EDCs are polar, non-volatile and thermally-labile compounds, LC is usually the best choice. For example, LC with UV detection has been used for the determination of BPA and APs [93], surfactants [99], PAHs and PCBs [47, 50, 53, 63], pharmaceuticals [110] and PBs [117]. LC with fluorimetric detection has not been widely used because the majority of studied pollutants do not exhibit native fluorescence. However, LC was used for the determination of several EDCs (BPA, APs and estrogens) [96], PAHs [53, 58], LAS and insoluble soap [105], pharmaceuticals [112] and PCBs [121] in sediments.

Recent advances in analytical instrumentation have allowed the unequivocal identification and confirmation of compounds at very low levels using LC-MS/MS. This technique has been used for the determination of pharmaceuticals [69, 74], UV-filters [73] and different surfactants [98]. LC-ESI-MS has been widely applied in the determination of a number of pollutants in marine sediments [100-104, 106-109, 111, 114, 116,117, 120, 136]. Finally, CE-DAD has seldom been used. Only a research work for the determination of TBBPA, TCBPA and related phenolic compounds was found in the literature [122].

4. Analytical Methods for Pollutants Determination in Biota

Until recently, the chemical analysis of contaminants in marine environments was restricted to monitoring water and sediment pollution. However, this is an incomplete approach to determine the whole environmental impact because the available data do not fully reflect the effects of the pollutants on aquatic organisms. Physicochemical behavior of contaminants depends on the particular affinity of these substances for the matrices, so that the existing concentrations in waters or sediments reflect only a partial view of the level of chemical contamination in the marine environment.

Therefore, a basic step in the impact assessment of pollutants in aquatic organisms involves establishing and validating analytical methods for their determination in biological matrices. The information about the environmental fate [12, 142], potential bioaccumulation in biota and biomagnification in the food chain of those pollutants is quantitatively lower [143]. Furthermore it is less known about the long-term effects of exposure to low levels of contaminants on wildlife and, ultimately, on humans [10].

Several studies have demonstrated the occurrence of micropollutants in marine waters and sediments, making them bioavailable to be taken up by filter-feeding organisms including mussels and oysters [119, 144-148, 150]. In addition, consumption of contaminated seafood represents a major source of human exposure to pollutants [151, 152].

Biota samples are complex matrices demanding extensive extraction and clean-up procedures to obtain extracts suitable for analysis. In addition, relatively low concentrations (ng g^{-1} or μg g^{-1}) may be expected to occur in these matrices. As a result, the analytical determination of the pollutants in marine organisms has been few studied and there is an increasing demand for reliable analytical methods [153].

It is worth mentioning pharmaceuticals in food matrices are subject to the requirements of food safety legislation [154-156]. In many cases, the methods developed for food matrices can be easily adapted to the analysis of biota, especially those developed for determination in fish, one of the most studied organisms. An overview of the most recently published methods for the analysis of pharmaceutical residues in biotic matrices has been provided by Huerta et al. [157].

4.1. Sample Treatment: Pretreatment, Extraction and Clean-Up

Considering the extreme complexity of biological matrices and the large differences in physical and chemical properties (e.g., polarity, solubility, stability) between groups of contaminants, very few multiclass methods have been developed for the screening of biota. In this respect, the paper of Ramirez et al. on the determination of 23 pharmaceuticals of different families in fish tissue is one of the most representative [158]. Most of the reported methods focus on the determination of single compounds or a family of compounds. In addition, the complexity of the matrix is a critical obstacle to overcome, since this complex matrix is especially rich in undesirable components that could interfere with the analysis resulting in low LOQs.

For liquid biological matrices, such as bile and blood, the preferred procedure is the dilution of sample to reduce matrix interferences, followed by preconcentration using SPE cartridges [159-162]. For compounds with acid-base properties, the pH should be adjusted to ensure the highest recoveries, as in the extraction of the β-blocker propranolol in plasma [163]. As β-blockers are basic compounds, they are predominantly in the non-ionized form in alkaline conditions and can be efficiently extracted using EtAc, with recoveries from 98 to 103 %. By adjusting pH, the extraction becomes selective, because this avoids the extraction of water-soluble components that could be responsible for matrix effects. Similarly, the analysis of ibuprofen in plasma samples [164] involves direct extraction with cold Ace and with Hex:EtAc (1:1, v/v). The use of cold organic solvents avoids the extraction of highly water-soluble components and offers selective determination of the target organic analytes.

For tissues and internal organs (e.g. muscle, liver), aggressive procedures are usually applied due to the strong interactions between analytes and matrix. However, LLE followed by conventional SPE clean-up provides good results, as in the determination of several NSAIDs and β-blockers in blue mussels [165]. Forty PPCPs were determined in fish fillet and liver tissue [166]. Residues of the perfluorinated chemicals PFOS and PFOA were determined in fish using salting-out liquid-liquid extraction [167].

In general, determination of POPs, as organochlorine compounds (PCBs, HCH isomers, HCB, DDT and chlordane) is done by simple analytical methods, based on known solid-liquid and LLE procedures [168], unlike emerging contaminants, that use more sophisticated extraction techniques.

USE is the simplest and most commonly used technique for the extraction of antibiotics, analgesics, antidepressants and hormones [169-172]. One of most important works developed a method for the analysis of 13 pharmaceuticals belonging to nine therapeutic classes in algal homogenates from microalga *Pseudokirchneriella subcapitata* [173]. The sample was extracted using several cycles of ultrasonic extractions with organic solvents, followed by SPE cleanup with cation exchange sorbents, due to acid-basic characteristics of the analytes. Traces of most of the target pharmaceuticals were detected in treated algal cells indicating that these algae are able to absorb pharmaceutical pollutants from the environment. Although DNA studies showed no genotoxic effect, proteomic analysis showed that algae, in particular the chloroplasts, are sensitive to the presence of drugs.

Other extraction method based on USE allowed the determination of barbital residues in fish tissue and shrimp muscle, but also in pork, poultry meat and swine liver [174]. USE extraction was followed by a clean-up with hexane and then a conventional SPE procedure was applied. Detection limits were 30 to100-fold lower than previous methods, showing that simple sample handling can serve as an attractive alternative analytical method for regulatory purposes.

Simpler sample treatments involving homogenization of sample with a solvent followed by a centrifugation step for macromolecules removal are also applied. This extraction procedure was used, for example, for the analysis of 24 pharmaceuticals in fish tissue [175].

Pharmaceuticals, xenoestrogens, 4-tert-octylphenol and bisphenol A were analyzed in fish tissue using MAE [176]. MAE can be improved using micellar systems. MAME have been applied to the detection of six pharmaceuticals in mollusk samples (mussels, clams, donax clams and oysters), using a non-ionic surfactant, polyoxyethylene 10 lauryl ether, coupled with SPE clean-up for the elimination of coeluents, with recoveries higher than 80% [177]. A similar technique based on enzymatic-MAE was developed for the determination of 11 antibiotics of five families and some of their main metabolites in fish tissue and mussel samples [178].

Recently, PLE has become a preferred technique, involving extraction with solvents at high pressure and temperature without reaching their critical point [179]. A method was proposed for extraction of macrolide antibiotics in fish muscle and meat, even though these compounds are especially temperature sensitive. Other types of pharmaceuticals, such as NSAIDs, lipid regulators, and antiepileptic drugs have been analyzed in marine bivalves [180]. This is an important analytical contribution, since it has been proved that these organisms are susceptible to exposure to pharmaceuticals.

Regarding to polar pesticides, an analytical procedure has been used for the detection of 29 pesticides in clams and seaweed using two alternative techniques: USE and PLE [181]. The results showed that PLE is a short-time consuming quantitative technique that requires smaller initial sample sizes. Greater specificity and selectivity in extraction and increased potential for automation were shown.

Recent studies have shown that selective serotonin reuptake inhibitors (SSRIs) accumulate in fish tissues. Residues of this group of psychiatric drugs and metabolites were also determined by PLE, followed by clean-up on SPE cartridges [182]. Paroxetine, fluoxetine and norfluoxetine were detected in samples.

Nevertheless, a common drawback of PLE is the coelution of matrix components, such as lipids (5-15 % in fish homogenates), pigments or proteins, must be removed to avoid analytical interferences. The simpler alternative is the clean-up of the extract with DCM using LLE, fatty components are retained in DCM which is further discarded. This procedure has been applied in the determination of BPA in fish [183].

Adsorption columns and GPC have also been used. The adsorption columns applied are usually polar sorbents in the normal phase for the removal of nonpolar lipids and other non-polar molecules. Florisil, alumina, silica gel, XAD-2, or aminosilane columns have often been used, especially for fatty samples, and sometimes mixed with an acid, alkaline or oxidizing reagent.

Florisil SPE cartridges were used as clean-up method for the analysis of ibuprofen and metabolites in different fish tissues [164]. GPC has the advantage that the column can be used over several months with no effects on clean-up capacity [179]. This technique has often been used in combination with other clean-up strategies, such as SPE, LLE and freezing the sample extracts in acetonitrile to precipitate remaining lipids such as cholesterol [184, 185]. This technique has been applied to the extraction of pharmaceuticals, PCPs and steroidal estrogens in fish tissue. A related technique that uses GPC after extraction and clean-up was performed on silica gel columns for the determination of five UV-filters (octocrylene, benzophenone-3, 4-methylbenzylidene camphor, 4-methylbenzylidene camphor and ethylhexylmethoxy cinnamate) in fish [186, 187].

Size exclusion chromatography (SEC) has also been applied [188-190], and in relatively few applications super critical CO_2 has been used as a mobile phase for eluting the analytes from the clean-up column [191-193].

The ability to rapidly determine bioaccumulation of these contaminants in exposed fish without sacrificing the animal (i.e. *in vivo*) would significantly facilitate research, assessment and monitoring of their environmental risk. For this purpose, and *in vivo* SPME approach was developed and applied to the measurement of a variety of emerging contaminants in fish. The results suggested that *in vivo* SPME is a potential alternative extraction technique for quantitative determination of contaminants in laboratory exposures and after exposure to municipal wastewater effluents, with the advantage of being a non-lethal tissue-sampling technique. Besides, SPME has additional advantages such as simplicity, sensitivity and robustness in these *in vivo* sampling conditions. For most of the compounds the results were comparable to those obtained using solid-liquid extraction to determine the total (extractable) concentrations of analytes in fish muscle [194, 195].

Other extraction techniques include ion-pair extraction and alkali-digestion for the determination of perfluorinated compounds in mussels, oysters, fish and dolphin muscle [196, 197].

In spite of the recognized advantages of the above mentioned techniques in comparison with Soxhlet extraction, some persistent organic pollutants like PCBs and DDTs can be determined using this technique, according to US EPA 1668a, which has been applied to many varieties of fish [198].

Finally, for typical determination of heavy metals in biological matrices, the fresh tissue and shell samples undergo a very similar treatment involving washing, hashing, drying at 105°C and mineralization using a wet digestion method (HNO_3-H_2SO_4), and analysis with an atomic absorption spectrophotometer in its several forms [199].

4.2. Analytical Techniques

In addition to the most commonly used analytical methods for identification and determination of contaminants in biota described in the literature, involving LC or GC in combination with MS detection, there are alternative techniques that have also been described and applied.

Less-known techniques, based on immunoassays have recently been developed for determination of antibiotics (e.g., sulfonamides and tetracyclines) and other pharmaceuticals in food products [200]. In particular, enzyme-linked immunosorbent assays, which provide antibiotic screening methods to determine residue levels of antibiotics in compliance with the current legislation.

The most commonly used biological element for the detection of veterinary drug residues is the antibody-antigen affinity, widely used as an immunochemical method. ELISA (and other labeled immunoassays) has been used to detect food contaminants for over 30 years. They are simple and inexpensive but slow methods, because they require multiple washing and incubation steps required and, unlike in medical applications, they have proved very hard to automate. Good repeatability has been difficult to achieve with ELISA. In many cases, such assays are considered to be only qualitative or, at best, semi-quantitative, and extraneous material in the sample can interfere with the detection system [201]. To overcome some of these problems, advanced techniques (e.g., label-free biosensor assays) have been developed over the past 10-15 years. Biosensors can provide real-time measurements, a high degree of automation and improved throughput and sensitivity. They are less expensive than sophisticated physicochemical instrumentation, require less time for analysis, are sensitive, automated and are more user-friendly. Since the advent of Biacore surface-plasmon resonance instruments in the early 1990s, a steadily growing use of this technology has been reported for the detection of food contaminants (e.g., veterinary drugs, mycotoxins, marine toxins, food dyes and processing contaminants). The use of other biosensing technologies (e.g., electrochemical and piezoelectric) also employed for the analysis of small-molecule contaminants may be on the rise. These techniques have already been developed for a wide range of antimicrobial compounds.

For β-lactams, some of the antibiotics most frequently used in Europe for the treatment of infections in animals, the challenge is not only to minimize preparation, extraction, clean-up and separation steps for the analysis in these complex matrices, but also to overcome the special physicochemical properties and instability of these compounds. For these reasons, some microbiological approaches in the form of rapid screening tests, usually using *B. Stearothermophilus*, var. *calidolactis*, have been developed. Although this methodology allows rapid screening of samples, it exhibits limited specificity, allowing only structurally similar compounds to be detected qualitatively as a group. Consequently, there is an increasing need for new methods that are also specific, besides sensitive, accurate, fast and simple [201]. Various screening methods to detect antibiotics have been developed that are based on an indirect competitive ELISA with photometric detection of horseradish peroxidase (HRP) activity, to detect oxytetracycline [203], and peroxidase-labeled goat anti-rabbit immunoglobulins to detect sulfonamide residues [204] in Gilthead sea bream (*Sparus aurata*) samples. Time-resolved fluoroimmunoassay (TR-FIA) using a secondary antibody with coproporphyrin of platinum (II) (PtCP) as a marker is also used [203]. The LOD obtained in fish extract with both techniques were suitable for fish quality control according to the maximum residue level established by the EU, although these techniques only provide semi-quantitative results.

A technique involving an optical biosensor inhibition immunoassay, based on the surface plasmon resonance principle, as a screening test for 13 (fluoro)quinolones for veterinary use in food-producing animals [205]. A number of quinolone derivatives were immobilised on the sensor chip and the binding of a range of different antibodies (polyclonal and one engineered antibody) was tested. This study represents an innovation in the production of generic polyclonal antibodies able to recognize several compounds with different chemical structures. For the first time, the concept of bi-active antibody was experimentally used and its efficiency was proved. The assay was applied to different biological matrices (poultry muscle, egg and fish). This rapid and simple method proved to be adequate for detecting (fluoro)quinolones at concentrations below the established maximum residue levels.

Despite the fact that the use of these biological is still very limited for the determination of pharmaceuticals, they have many advantages including high sensitivity, simplicity, and cost-effectiveness, which makes them good screening techniques for different pharmaceutical classes [202].

Regarding instrumental techniques, LC in tandem with spectrophotometric detection, such as diode-array detection, UV absorbance detection and fluorescence detection, has been used extensively for the detection of some pharmaceutical classes in aquatic organisms. Some studies for the determination of quinolones in fish, crustaceans, and aquatic plants were performed using LC-FLD [206-209], UV absorbance was used for the detection of NSAIDs, lipid regulators and anti-epileptic drugs in mussel tissue [177], with very low LOD. The use of these techniques is, however, limited to compounds with specific physicochemical characteristic, such as the presence of chromophores or fluorescent properties. Although LC with diode-array detection, UV absorbance detection, or fluorescence detection are proved may be a cost-effective technology, the coupling of chromatography with MS detection is the method of choice for the determination of all classes of contaminants in environmental matrices, given its high selectivity, specificity, and sensitivity [9, 210, 211, 212]. LC-MS and LC-MS/MS have gained great popularity in the last decade owing to their compatibility with polar and non-volatile compounds. Most of the analytical methods developed for the analysis of contaminants in biota are based on ESI interface [172, 213-218]. APCI has been used for the detection of antidepressants in fish homogenates, using a standard addition method for correction of possible matrix effects [182]. A growing trend in chromatography is the use of ultra high performance liquid chromatography (UHPLC). UHPLC is a new technique that uses small-diameter particles in the stationary phase and short columns. This technique has been used for the determination of 13 different aminoglycoside antibiotics in fish muscle, veal liver and kidney [219] and diclofenac in bile of rainbow trout (*Oncorhynchus mykiss*) [220].

GC coupled either with MS or ECD and capillary GC-MS are the methods of choice for compounds like PAHs, PCBs, and OCPs [221]. Capillary GC-MS has been applied for the detection of antidepressants in several fish tissues including fatty samples from the liver, using pentafluoropropionic anhydride as derivatizing reagent, with excellent LOD [215].

Both GC-MS and LC-MS may be affected by matrix interferences, which can reduce the sensitivity and increase considerably the LODs of the analytical methods. Matrix effects are particularly strong at low analyte concentration, which is a common situation for contaminants in biota [222, 223], ranging from 56% suppression to 25% enhancement in the analysis of benthic invertebrates [221]. Although there is a clear impact of matrix on ionization of PPCPs, matrix effects can be quantified and overcome. Different approaches have been used to overcome matrix effects. The most popular one consists in the addition of labeled compounds as internal standards, which can correct variations during instrumental detection. Other corrective measures are matrix-matched calibration and standard addition, but they also have their drawbacks including the need for an uncontaminated matrix for the preparation of a matrix-matched calibration that is usually difficult to obtain, and having to generate a calibration curve for each sample that becomes impractical when analyzing a large number of samples.

5. Biomonitoring and Ecological Risk Assessment for Emerging Pollutants

Biomonitoring tools (e.g. bioassays, biomarkers, microbial community analyses) have a great potential for increasing reliability in the risk assessment of both regulated and emerging contaminants. Sensors developed to determine several analytes simultaneously are useful tools in environmental monitoring and screening. The comprehensive term 'biosensor' refers to a system capable of detecting the presence of a substrate that uses biological components and that provides a quantifiable signal [224].

Biomonitoring is a scientific technique for environmental assessment, including human exposure to natural and anthropogenic compounds, based on sampling and analysis of an individual organism's (biomarkers/bioindicators) tissues and fluids. Different species can be used as indicators for the surveillance of pollutants uptake, bioavailability, excretion and study of toxic effects. The obtained information will provide a complete view about the level of environmental pollution.

Recently, research on biosensing techniques and devices for environmental monitoring have included genetic engineering and sensor cell development.

For example, it is assumed that many endocrine disruptors can bind to the estrogen receptor (ER) as agonists or antagonists. Therefore, the chemical binding capacity of the ER would be a factor to be determined in screening or testing the potential environmental toxicity of these substances. The development of biosensors for endocrine disruptors will take advantage of this property [225]. In addition, molecular self-assembly, inspired by nature, has been proposed to synthesize nanostructures with distinctive functions, because current detection methods for pathogenic bacteria, protozoa, viruses, or helminths proved to be sometimes inaccurate and costly and time-consuming.

The self-assembly technique makes possible to create organized, patterned nanostructures with biomaterials (proteins, lipids, nucleic acids), without external control and directions, used in the development of amperometric immunosensors [226].

Genomics has become a new tool in recognizing and understanding the molecular pathways disturbed by emerging pollutants and is able to relate them to both the whole organism and population level effects. Technologies such as DNA microarrays are successfully applied in ecotoxicogenomics, an emerging field in ecotoxicology, to understand the effects of pollutants at the molecular level [227]. (Bio)monitoring provides information about concentration levels of pollutants in a different way than analytical techniques. (Bio)monitoring allows predicting microbial degradation and environmental availability of contaminants [228]. Consequently, it is crucial to develop techniques able to predict microbial bioavailability and environmental degradation of pollutants. Quantifying the bioavailability of organic contaminants in soil and sediment is crucial for risk assessment and decision making in contaminated soils and sediments remediation.

6. Emerging Technologies in Bioremediation

Bioremediation offers an environmentally friendly and cost-effective option to remove contaminants from the environment. Because bioremediation exploits the natural ability of plants and microbes for decontamination, theoretically organic pollutants could be completely mineralized into water and carbon dioxide. Thanks to biotechnological alternatives some of these pollutants can be degraded or removed. Studies have demonstrated that these solutions involving microbes, plants and animals under specific conditions that address both abiotic and biotic factors, can achieve contaminant mineralization, transformation or immobilization [229]. Bioremediation strategies seem to cause less environmental impact and are more cost-effectively than conventional techniques [230]. In addition to this, bioremediation is a non-invasive technique that requires lower concentration of contaminants with which the clean-up by physical or chemical methods would not be feasible. Unfortunately, bioremediation has some major drawbacks that limit implementation, including the long processes involved and the low predictability. The strategies for bioremediation are the following [229, 230]:

A. *Monitored natural recovery (MNR)* uses the "self-healing" natural capacities of the indigenous microbial population in combination with natural occurring physical and chemical processes.
B. *Biostimulation* encourages the indigenous population by influencing the factors affecting microbial growth. This, means stimulating the degradation pathways of contaminants by addition of required substances or providing favorable conditions. For example, as anaerobic sediment conditions are necessary for biodegradation of some contaminants, and anaerobic metabolic pathways need electron acceptors other than oxygen, stimulation is achieved by adding these acceptors. Studies on benzene degradation in sediments from petroleum-contaminated aquifers showed anaerobic benzene degradation when electron acceptors like Fe (III), sulfate, nitrate and humic substances were available [231].

The biodegradation capacity of a compound depends on its bioavailability, which at the same time is related to physical and chemical properties of the substances. Surfactants are used as potential agents for enhanced solubility and removal of contaminants from sediments [232, 233]. Biosurfactants, like bile salt, sodium taurocholate, and synthetic surfactants, like Triton X-100 and sodium dodecyl sulphate, have proved to increase the bioavailability of contaminants like PAHs and PCBs [234, 235]. According to several studies, the use of surfactants, of biological and synthetic origin, could result in between facilitating and retarding biodegradation or not affecting biodegradation at all. For example, hydrocarbon degradation is affected by biosurfactants because these increase the solubility of the contaminant in the aqueous phase and change the affinity between the microbial cells and the hydrocarbons by increasing cell surface hydrophobicity [236]. But surfactants do not always facilitate biodegradation, since they can also inhibit it by detaching the cells from the contaminant-water interface [237], when the surfactant itself turns to be substrate to degrading microorganisms [238] or when the surfactant causes a toxic effect on the biodegrading microorganisms [239, 240]. It is important to take into account the adsorption capacity of the surfactant to soil or sediment, since the amount of surfactants adsorbed decreases the concentration available for dissolving the contaminant [241].

C. *Bioaugmentation.* Introducing appropriate species for the degradation of specific contaminants. Microorganisms with specific catabolic abilities are introduced into the contaminated environment in order to supplement the indigenous population and to speed up or enable the degradation of pollutants. Bioaugmentation has successfully been used for remediation of PAHs in sediments with poor or lacking intrinsic degradation potential [242, 243] while other studies demonstrated that bioaugmentation did not enhance biodegradation significantly compared to natural attenuation [244].

One of the main problems in applying bioaugmentation is to ensure the survival and activity of the introduced organisms in the environment. The key factor to successful bioaugmentation is the selection of the appropriate bacterial strain and taking into account the indigenous microbial communities in the source habitat [245]. The selection of an adequate strain for remediation purposes depends not only on its degradation ability, but also on its ecological characteristics concerning adaptation to the habitat. In a series of experiments aimed to develop a bacterial inocula to treat spent metal working fluids in bioreactors, three criteria where used to select strains: *(i)* the relative abundance of the source populations in the target habitat (waste), *(ii)* tolerance to co-contaminants and *(iii)* the ability to degrade target contaminants [246]. The final consortium of four strains was 85% more effective at processing the waste than undefined inocula from sewage. Its performance was also more consistent and predictable [247].

Choosing bioaugmentation as remediation strategy is viable if the limiting factor of biodegradation is the absence of relevant catabolic genes within the indigenous microbial community and this lack of genetic information will be filled by the introduced strain [245]. Genetic information containing degradation abilities may also be introduced by donor species, which themselves do not survive a long time in the environment but pass on the lacking genetic information to intrinsic microorganisms [248, 249].

D. *Phytoremediation* involves the use of plants and algae in the degradation and removal of contaminants from the environment. This technique has the potential to completely mineralize or transform the pollutant into a less or non-toxic component. Phytoremediation has been used to remove PCBs, PAHs, nitroaromatics and halogenated hydrocarbons. The mechanisms of phytoremediation include biophysical and biochemical processes like adsorption, transport and translocation, as well as transformation and mineralization by plant enzymes [250]. Plants have been shown to be able to degrade halogenated compounds like trichloroethene by oxidative degradation pathways, including plant specific dehalogenases [251], whose activity was observed to be maintained after the plants death, enzymes are able to bind to the organic matrix of the sediment as plants die, then they are buried in the sediment, contributing to the dehalogenase activity observed in organic-rich sediments [252]. A disadvantage of phytoremediation is that detoxification of organic pollutants is often slow and if decomposition is not complete, toxic compounds may accumulate in plant tissue and can be released to the environment or enter food-chains [253].

Phytoremediation has been investigated for remediation of marine sediments. In a recent study, benthic microalgae have been used for the first time in the remediation of organically enriched sediments [254]. Intrinsic benthic microalgae species were mass cultured in the laboratory and reintroduced to the environment attached to glass beads. Re-plantation changed the reduced conditions of the organic-rich sediment to toxic, thus enhancing the degradation of the organic material.

Moreover, phytoremediation can be applied in combination with other bioremediation strategies, e.g. biostimulation, which could improve the efficiency of decontamination strategy.

CONCLUSION

Although the understanding of the effects and distribution of the different types of pollutants has increased substantially since the early reports on the biological effects of marine pollution and associated monitoring problems, there is still a need to develop methods for the identification, estimation, comparative assessment and management of the potential risks posed by chemical pollutants to aquatic living resources and marine ecosystems. In addition, it is necessary to develop and validate analytical tools capable of detecting these compounds in different compartments of the marine environment. This will enable to obtain quantitative data that will help to know the fate of these chemicals, in order to prevent, limit or reduce the environmental impact. The development of accurate analytical methods to determine pollutants in marine enviroments has become an issue of mayor interest and as a result a large number of studies focusing on the determination of different classes of pollutants in seawater have been recently published. Regarding experimental issues, due to the low concentration levels of pollutants in this environment, the optimization of new sample treatment procedures is paramount. In this way, a sample clean-up to remove interferences and concentrate the studied pollutants are required for a selective and sensitive determination.

In this context, an increasing number of studies demonstrate the presence of a large amount of pollutants (PAHs, PCBs, PBDEs, OCPs, surfactants, PCPs, EDCs and heavy metals) in marine environments. Multiresidue methods are being developed for the determination of several families of pollutants with one extraction step and limited sample preparation, which is one of the most critical stages of the analytical procedure. In addition to the traditional LLE and SPE, microextraction and intrumental extraction techniques (USE, MAE, PLE) are becoming alternatives for the analysis of marine solid samples. Regarding analytical techniques, GC-MS has been the most used technique for quantifying PAHs, PCBs, PBDEs and OCPs, being GC-ECD an alternative for halogenated compounds. On the other hand, non-volatile compounds such as most PPCPs and many EDCs have usually been analyzed with LC-UV, LC-FLD or LC-MS/MS. At the same time, some biomonitoring tools for the detection of a number of pollutants in marine media have been developed.

Finally and thanks to all these developed techniques, it has been possible to know the extent of contamination in numerous marine environments resulting in the advent of numerous bioremediation strategies such as monitored natural recovery (MNR), biostimulation, bioaugmentation and phytoremediation.

REFERENCES

[1] Moore, MN; Depledge, MH; Readman, JW; Paul Leonard DR. An integrated biomarker–based strategy for ecotoxicological evaluation of risk in environmental management. *Mutation Research,* 2004, 552, 247–268.

[2] Meryl, W. The transition in the contribution of living aquatic resources to food security, in: Pap 13 of International Food Policy Research Institute, Washington D.C., 1996, 27–28.

[3] FAO, The State of World Fisheries and Aquaculture, 2009.

[4] Halpern, BS; Walbridge, S; Selkoe, KA; Kappel, CV; Micheli, F; D'Agrosa, C; Bruno, JF; Casey, KS; Ebert, C; Fox, HE; Fujita, R; Heinemann, D; Lenihan, HS; Madin, EMP; Perry, MT; Selig, ER;

Spalding, M; Steneck, R; Watson, R. A global map of human impact on marine ecosystems. *Science*, 2008, 319, 948–952.

[5] Lotze, HK; Lenihan, HS; Bourque, BJ; Bradbury, RH; Cooke, RG; Kay, MC; Kidwell, SM; Kirby, MX; Peterson, CH; Jackson, JBC. Depletion, degradation, and recovery potential of estuaries and coastal seas. *Science*, 2006, 312, 1806–1809.

[6] GESAMP, Scientific Criteria for the selection of Waste Disposal Sites at Sea. Reports and Studies No. 16, Inter–Governmental Maritime Consultative Organization, London, 1982.

[7] Marine Strategy Framework Directive (Directive 2008/56/EU). Date of adoption: 17 June 2008. Entry into force: 15 July 2008 (OJ L 164/19 25.6.2008). Deadline for transposition by Member States: 15 July 2010

[8] Binelli, A; Provini, A; The PCB pollution of Lake Iseo (N. Italy) and the role of biomagnification in the pelagic food web. *Chemosphere*, 2003, 53, 143–151.

[9] Holsbeek, L; Joiris, CR; Debacker, V; Ali, BI; Roose, P; Nellissen, JP; Gobert, S; Bouquegneau, JM; Bossicart, M. Heavy metals, organochlorines and polycyclic aromatic hydrocarbons in sperm whales stranded in the southern North Sea during the 1994/1995 winter. *Marine Pollution Bulletin,* 1999, 38, 304–313.

[10] Sarkar A. Occurrence and distribution of persistent chlorinated hydrocarbons in the seas around India, in: S. Majumdar, E. Miller, G. Forbes, R. Schmalz, A.A. Panah (Eds.), The oceans: physico–chemical dynamics and resources, The Pennsylvania Academy of Science, USA, 1994, 445–449.

[11] Solé, M; Porte, C; Albaigés J. Hydrocarbons, PCBs and DDT in the NW Mediterranean deep–sea fish Mora moro. *Deep Sea Research Part I: Oceanographic Research Papers,* 2001, 48, 495–513.

[12] Kucklick, J; Struntz, W; Becker, P; York, G; O'Hara, T; Jesssica, E; Persistent organochlorine pollutants in ringed seals and polar bears collected from northern Alaska, *Science of the Total Environment,* 2002, 287, 45–59.

[13] Shore, R; Wright, J; Horne, J; Sparks, T. Polycyclic aromatic hydrocarbon (PAH) residues in the eggs of coastal–nesting birds from Britain. *Marine Pollution Bulletin,* 1999, 38, 509–513.

[14] Carajaville, M; Bebianno, M; Blasco, J; Porte, C; Sarasquete, C; Viarengo, A. The use of biomarkers to assess the impact of pollution in coastal environment of the Iberian Peninsula: a practical approach, *Science of the Total Environment,* 2000, 247, 295–311.

[15] Mostofa, KMG; Liu, CQ; Mottaleb, A; Wan, GJ; Ogawa, H; Vione, D; Yoshioka, T; Wu, FC. Dissolved organic matter in natural waters, in: Mostofa, KMG; Yoshioka, T; Mottaleb, A; Vione, D (Eds)), Photobiogeochemistry of organic matter: principles and practices in water environment, Springer, New York, 2013, 1–137.

[16] Haynes, D; Johnson, JE. Organochlorine, heavy metal and polyaromatic hydrocarbon pollutant concentrations in the Great Barrier Reef (Australia) environment: a review. *Marine Pollution Bulletin,* 2000, 41, 7–12.

[17] Pinto, E; Sigaud–Kutner, TCS; Leittao, MAS; Okamoto, OK; Morse, D; Colepicolo, P. Heavy metal–induced oxidative stress in algae, *Journal of Phycology*, 2003, 39, 1008–1018.

[18] De Laender, F; Hammer, J; Hendriks, AJ; Soetaert, K; Janssen, CR. Combining monitoring data and modeling identifies PAHs as emerging contaminants in the Arctic. *Environmental Science and Technology*, 2011, 45, 9024–9029.

[19] Richardson, S.D; Ternes, T.A; Water analysis: emerging contaminants and current issues, *Analytical Chemistry,* 2011, 83, 4614–4648.

[20] Hogenboom, AC; van Leerdam, JA; de Voogt, P. Accurate mass screening and identification of emerging contaminants in environmental samples by liquid chromatography–hybrid linear ion trap orbitrap mass spectrometry. *Journal of Chromatography A,* 2009, 1216, 510–519.

[21] Vidal–Dorsch, DE; Bay, SM; Maruya, K; Snyder, SA; Trenholm, RA; Vanderford BJ. Contaminants of emerging concern in municipal wastewater effluents and marine receivingwater. *Enviromental Toxicology and Chemistry,* 2012, 31, 2674–2682.

[22] Witt, G. Polycyclic aromatic hydrocarbons in water and sediments of the Baltic Sea. *Marine Pollution Bulletin*, 1995, 31, 237–248.

[23] Kolpin, DW; Furlong, ET; Meyer, MT; Thurman, EM; Zaugg, SD; Barber, LB; Buxton, HT. Pharmaceuticals, hormones, and other organic wastewater contaminants in U.S. streams, 99–00: A national reconnaissance. *Environmental Science and Technology,* 2002, 36, 1202–1211.

[24] Corcoran, J; Winter, MJ; Tyler, CR. Pharmaceuticals in the aquatic environment: a critical review of the evidence for health effects in fish. *Critical Reviews in Toxicology,* 2010, 40, 287–304.

[25] Santos, LHMLM; Araújo, AN; Fachini, A; Pena, A; Delerue–Matos, C; Montenegro MCBSM. Ecotoxicological aspects related to the presence

of pharmaceuticals in the aquatic environment. *Journal of Hazardous Materials,* 2010, 175, 45–95.

[26] Crane, M; Watts, C; Boucard, T. Chronic aquatic environmental risks from exposure to human pharmaceuticals. *Science of the Total Environment,* 2006, 367, 27–41.

[27] Cuthbert, R; Taggart, M.A; Prakash, V; Saini, M; Swarup, D; Upreti, S; Mateo, R; Chakraborty, SS; Dori, P; Green RE. Effectiveness of action in India to reduce exposure of Gyps Vultures to the toxic veterinary drug diclofenac. *PLoS One,* 2011, 65, e19069.

[28] Jobling, S; Coey, S; Whitmore, J.G; Kime, D.E; Van Look, K.J.W; McAllister, B.G; Beresford, N; Henshaw, AC; Brighty, G; Tyler, CR; Sumpter, JP. Wild intersex roach (*Rutilus rutilus*) have reduced fertility. *Biology of Reproduction,* 2002, 67, 515–524.

[29] Brooks, B.W; Chambliss, C.K; Stanley, J.K; Ramirez, A; Banks, K.E; Johnson, R.D; Lewis, RJ. Determination of select antidepressants in fish from an effluent dominated stream. *Enviromental Toxicology and Chemistry,* 2005, 24, 464–469.

[30] Nassef, M; Matsumoto, S; Seki, M; Khalil, F; Kang, IJ; Shimasaki, Y; Oshima, Y; Honjo, S. Acute effects of triclosan, diclofenac and carbamazepine on feeding performance of Japanese medaka fish *Oryzias latipes*. *Chemosphere,* 2010, 80, 1095–1100.

[31] Thomulka, KW; McGee DJ. Detection of biohazardous materials inwater by measuring bioluminescence reduction with the marine organism *Vibrio harveyi*. *Journal of Environmental Science and Health,* Part A, 1993, 28, 219–227.

[32] Pro, J; Ortiz, JA; Boleas, S; Fernandez, C; Carbonell, G; Tarazona, JV. Effect assessment of antimicrobial pharmaceuticals on the aquatic plant Lemna minor. Bulletin *of Environmental Contamination and Toxicology,* 2003, 70, 290–295.

[33] Yamashita, N; Yasojima, M; Miyajima, K; Suzuki, Y; Tanaka, H. Effects of antibacterial agents, levofloxacin and clarithromycin, on aquatic organisms. *Water Science and Technology,* 2006, 53, 65–72.

[34] Constanzo, SD; Murby, J; Bates, J. Ecosystem response to antibiotics entering the aquatic environment. *Marine Pollution Bulletin,* 2004, 51, 218–233.

[35] Fent, K; Weston, AA; Caminada, D. Ecotoxicology of human pharmaceuticals. *Aquatic Toxicology,* 2006, 76, 122–159.

[36] EC 2000. Directive 2000/60/EC of the European Parliament and of the council establishing a framework for community action in the field of

wáter policy. *Official Journal of the European Comunities*, L327, European Communities, Brussels.

[37] Hirsch, R; Ternes, T; Haberer, K; Kratz, KL. Occurrence of antibiotics in the aquatic environment. *Science of the Total Environment,* 1999, 225, 109–118.

[38] Mostofa, KMG; Liu, C; Mottaleb, MA; Wan, G; Photobiogeochemistry of Organic Matter. Springer Berlin Heidelberg, Berlin, Heidelberg, 2013.

[39] De Mora, S; Fowler, SW; Wyse, E; Azemard, S. Distribution of heavy metals in marine bivalves; fish and coastal sediments in the Gulf and Gulf of Oman. *Marine Pollution Bulletin,* 2004, 49, 410–424.

[40] Saito, MA; Schneidera, DL. Examination of precipitation chemistry and improvements in precision using the $Mg(OH)_2$ preconcentration inductively coupled plasma mass spectrometry (ICP–MS) method for high–throughput analysis of open–ocean Fe and Mn in seawater. *Analytica Chimica Acta,* 2006, 565, 222–233.

[41] Ellwood MJ. Wintertime trace metal (Zn, Cu, Ni, Cd, Pb and Co) and nutrient distributions in the Subantarctic Zone between 40–52° S; 155–160°E. *Marine Chemistry,* 2008, 112, 107–117.

[42] de Jonga, J; Schoemann, V; Lannuzel, D; Tisond, JL; Mattiellia, N. High–accuracy determination of iron in seawater by isotope dilution multiple collector inductively coupled plasma mass spectrometry (ID–MC–ICP–MS) using nitrilotriacetic acid chelating resin for pre–concentration and matrix separation. *Analytica Chimica Acta*, 2008, 623, 126–139.

[43] Ghanemi, K; Nikpour, Y; Omidvar, O; Maryamabadi, A. Sulfur–nanoparticle–based method for separation and preconcentration of some heavy metals in marine samples prior to flame atomic absorption spectrometry determination. *Talanta,* 2011, 85, 763–769.

[44] Martinis, EM; Olsina, RA; Altamirano, JC; Wuilloud, RG. Sensitive determination of cadmium in water samples by room temperature ionic liquid–based preconcentration and electrothermal atomic absorption spectrometry. *Analytica Chimica Acta,* 2008, 628, 41–48.

[45] He, Q; Hu, Z; Jiang, Y; Chang, X; Tu, Z; Zhang, L. Preconcentration of Cu(II); Fe(III) and Pb(II) with m2–((2–aminoethylamino)methyl)phenol–functionalized activated carbon followed by ICP–OES determination. *Journal of Hazardous Materials,* 2010, 175, 710–714.

[46] Rahmi, D; Takasaki, Y; Zhu, Y; Kobayashi, H; Konagaya, S; Haraguchi, H; Umemur, T. Preparation of monolithic chelating adsorbent inside a syringe filter tip for solid phase microextraction of trace elements in natural water prior to their determination by ICP–MS. *Talanta*, 2010, 81, 1438–1445.

[47] Kot–Wasik, A; Dezbska, J; Namiesnik, J. Monitoring of organic pollutants in coastal waters of the Gulf of Gdansk, Southern Baltic. *Marine Pollution Bulletin,* 2004, 49, 264–276.

[48] Qingling, L; Xiaoqin, X; Sen–Chun, LF; Xiaoru, W. Determination of trace PAHs in seawater and sediment pore–water by solid–phase microextraction (SPME) coupled with GC/MS. *Science China,* 2006, 49, 481–491.

[49] Pérez–Carrera, E; León, VM; Gómez, A; González–Mazo, E. Simultaneous determination of pesticides; polycyclic aromatic hydrocarbons and polychlorinated biphenyls in seawater and interstitial marine water samples; using stir bar sorptive extraction–thermal desorption–gas chromatography–mass spectrometry. *Journal of Chromatography A,* 2007, 1170, 82–90.

[50] Habibi, MH; Hadjmohammadi, MR. Determination of some polycyclic aromatic hydrocarbons in the Caspian Seawater by HPLC following preconcentration with Solid–Phase Extraction. *Iranian Journal of Chemistry and Chemical Engineering,* 2008, 27, 91–96.

[51] Prieto, A; Telleria, O; Etxebarria, N; Fernández, LA; Usobiaga, A; Zuloaga, O. Simultaneous preconcentration of a wide variety of organic pollutants in water samples. Comparison of stir bar sorptive extraction and membrane–assisted solvent extraction. *Journal of Chromatography A,* 2008, 1214, 1–10.

[52] Qiu, YW; Zhang, G; Liu, GQ; Guo, LL; Li, XD; Wai, O. Polycyclic aromatic hydrocarbons (PAHs) in the water column and sediment core of Deep Bay, South China. *Estuarine, Coastal and Shelf Science,* 2009, 83, 60–66.

[53] Stortini, AM; Martellini, T; Del Bubba, M; Lepri, L; Capodaglio, G; Cincinelli A. n–Alkanes; PAHs and surfactants in the sea surface microlayer and sea water samples of the Gerlache Inlet sea (Antarctica). *Microchemical Journal,* 2009, 92, 37–43.

[54] Grueiro–Noche, G; Andrade, JM; Muniategui–Lorenzo, S; López–Mahía, P; Prada–Rodríguez, D. 3–Way pattern–recognition of PAHs from Galicia (NW Spain) seawater samples after the Prestige's wreck. *Environmental Pollution,* 2010, 158, 207–214.

[55] March, JG; Moukhchan, F; Cerdà, V. Application of in–vial membrane assisted solvent extraction to the determination of polycyclic aromatic hydrocarbons in seawater by gas chromatography–mass spectrometry. *Analytica Chimica Acta,* 2011, 685, 132–137.

[56] Wu, YL; Wang, XH; Li, YY; Hong, HS. Occurrence of polycyclic aromatic hydrocarbons (PAHs) in seawater from the Western Taiwan Strait, China. *Marine Pollution Bulletin*, 2011, 63, 459–463.

[57] Song, X; Li, J; Liao, C; Chen, L. Ultrasound–assisted dispersive liquid–liquid microextraction combined with low solvent consumption for determination of polycyclic aromatic hydrocarbons in seawater by GC–MS. *Chromatographia,* 2011, 89–98.

[58] Li, Y; Yoshida, S; Chondo, Y; Nassar, H; Tang, N; Araki, Y; Toriba, A; Kameda, T; Hayakawa, K. On–line concentration and fluorescence determination HPLC for polycyclic aromatic hydrocarbons in seawater samples and its application to Japan Sea. *Chemical and Pharmaceutical Bulletin,* 2012, 60, 531–535.

[59] Kraleva, E; Karamfilov, V; Hibaum, G. Determination of PAH in the Black Sea water by GC/MS following preconcentration with solid phase extraction. *Ecological Chemistry and Engineering. S,* 2012, 19, 393–403.

[60] Mirza, R; Mohammadi, M; Sohrab, AD; Safahieh, A; Savari, A; Hajeb, P. Polycyclic aromatic hydrocarbons in seawater; sediment and rock oyster saccostrea cucullata from the Northern, Part of the Persian Gulf (Bushehr Province). *Water, Air and Soil Pollution,* 2012, 223, 189–198.

[61] Liu, X; Jia, H; Wang, L; Qi, H; Ma, W; Hong, W; Guo, J; Yang, M; Sun, Y; Li, YF. Characterization of polycyclic aromatichydrocarbon sin concurrently monitored surface seawater and sediment along Dalian coast after oil spill. *Ecotoxicology and Environmental Safety,* 2013, 90, 151–156.

[62] Rojo–Nieto, E; Salesb, D; Peralesa, JA. Sources; transport and fate of PAHs in sediments and superficial water of a chronically polluted semi–enclosed; body of seawater: linking of compartments. *Environmental Science,* 2013, 15, 986–995.

[63] Rima J. HPLC and spectrofluorimetric determination of pyrene in the Lebanese coast. *Journal of the Black Sea/Mediterranean Environment,* 2013, 1, 958–969.

[64] Ramirez, CE; Wang, C; Gardinali, PR. Fully automated trace level determination of parent and alkylated PAHs in environmental waters by

online SPE–LC–APPI–MS/MS. *Analytical and Bioanalytical Chemistry,* 2014, 406, 329–344.

[65] Babu Rajendrana, R; Imagawaa, T; Taoa, H; Ramesh, R. Distribution of PCBs; HCHs and DDTs, and their ecotoxicological implications in Bay of Bengal, India. *Environmental International,* 2005, 31, 503–512.

[66] Basheer, C; Vetrichelvan, M; Valiyaveettil, S; Lee, HK. On–site polymer–coated hollow fiber membrane microextraction and gas chromatography–mass spectrometry of polychlorinated biphenyls and polybrominated diphenyl ethers. *Journal of Chromatography A*, 2007, 1139, 157–164.

[67] Song, X; Li, J; Chen, L; Cai, Z; Liao, C; Peng, H, Xiong, H. Determination of polychlorinated biphenyls in seawater using headspace solid–phase microextraction coupled with gas chromatography–mass spectrometry with the aid of experimental design. *Journal of the Brazilian Chemical Society,* 2012, 23, 132–141.

[68] Hassan, J; Manavi, PN; Darabi, E. Polychlorinated biphenyls hot and cold seasons distribution in see water; sediment; and fish samples in the Khour–e–Mousa (Mah–Shahr), Iran. *Chemosphere,* 2013, 90, 2477–2482.

[69] Weigel, S; Berger, U; Jensen, E; Kallenborn, R; Thoresen, H; Huhnerfuss, H. Determination of selected pharmaceuticals and caffeine in sewage and seawater from Tromsø/Norway with emphasis on ibuprofen and its metabolites. *Chemosphere,* 2004, 56, 583–592.

[70] Wille, K; Noppe, H; Verheyden, K; Vanden, Bussche, J; De Wulf, E; Van Caeter, P; Janssen, CR; De Brabander, HF; Vanhaecke, L. Validation and application of an LC–MS/MS method for the simultaneous quantification of 13 pharmaceuticals in seawater. *Analytical and Bioanalytical Chemistry,* 2010, 397, 1797–1808.

[71] Magnér, J; Filipovic, M; Alsberg, T. Application of a novel solid–phase–extraction sampler and ultra–performance liquid chromatography quadrupole–time–of–flight mass spectrometryfor determination of pharmaceutical residues in surface sea water. *Chemosphere,* 80, 2010, 1255–1260.

[72] Tarazona, I; Chisvert, A; León, Z; Salvador, A. Determination of hydroxylated benzophenone UV filters in sea water samples by dispersive liquid–liquid microextraction followed by gas chromatography–mass spectrometry. *Journal of Chromatography A*, 2010, 1217, 4771–4778.

[73] Nguyen, KTN; Scapolla, C; Di Carro, M; Magi, E. Rapid and selective determination of UV filters in seawater by liquid chromatography–tandem mass spectrometry combined with stir bar sorptive extraction. *Talanta,* 2011, 85, 2375– 2384.

[74] Borecka, M; Białk–Bielinska, A; Siedlewicz, G; Kornowska, K; Kumirska, J; Stepnowski, P; Pazdro, K. A new approach for the estimation of expanded uncertainty of results of an analytical method developed for determining antibiotics in seawater using solid–phase extraction disks and liquidchromatography coupled with tandem mass spectrometry technique. *Journal of Chromatography A,* 2013, 1304, 138–146.

[75] Benedé, JL; Chisvert, A; Salvador, A; Sánchez–Quiles, D; Tovar–Sánchez, A. Determination of UV filters in both soluble and particulate fractions ofseawaters by dispersive liquid–liquid microextraction followed by gaschromatography–mass spectrometry. *Analytica Chimica Acta,* 2014, 812, 50–58.

[76] Kotnik, K; Kosjek, T; Krajnc, U; Heath, E. Trace analysis of benzophenone–derived compounds in surface waters and sediments using solid–phase extraction and microwave–assisted extraction followed by gas chromatography–mass spectrometry. *Analytical and Bioanalytical Chemistry.* DOI 10.1007/s00216–014–7749–0.

[77] Sapota, G. Polychlorinated biphenyls (PCBs) and organochlorine pesticides (OCPs) in seawater of the southern Baltic Sea. *Desalination,* 2004, 162, 153–157.

[78] Kurt, PB; Ozkoc, HB. A survey to determine levels of chlorinated pesticides and PCBs in mussels and seawater from the Mid–Black Sea Coast of Turkey. *Marine Pollution Bulletin,* 2004, 48, 1076–1083.

[79] Wurl, O; Obbard, JP. Chlorinated pesticides and PCBs in the sea–Surface microlayer and seawater samples of Singapore. *Marine Pollution Bulletin,* 2005, 50, 1233–1243.

[80] Alawi, M; Rasheed, M; Al–Masri, MG. Organochlorine pesticides in seawater and sediment samples from the Aqaga Gulf. *Fresenius Environmental Bulletin,* 2007, 16, 1131–1136.

[81] Qiu, YW; Zhang, G; Guo, LL; Cheng, HR; Wang, W; Li, XD; Wai, OWH. Current status and historical trends of organochlorine pesticides in the ecosystem of Deep Bay, South China. *Estuarine, Coastal and Shelf Science,* 2009, 85, 265–272.

[82] Zhong, G; Xie, Z; Cai, M; Möller, A; Sturm, R; Tang, J; Zhang, G; He, J; Ebinghaus, R. Distribution and air–sea exchange of current–use

pesticides (CUPs) from East Asia to the High Arctic Ocean. *Environmental Science and Technology,* 2012, 46, 259–267.
[83] Lin, T; Li, J; Xu, Y; Liu, X; Luo, C; Cheng, H; Chen, Y; Zhang, G. Organochlorine pesticides in seawater and the surrounding atmosphere of the marginal seas of China: Spatial distribution; sources and air–water Exchange. *Science of the Total Environment,* 2012, 435/436, 244–252.
[84] Huang, Y; Xu, Y; Li, J; Xu, W; Zhang, G; Cheng, Z; Liu, J; Wang, Y; Tian, C. Organochlorine pesticides in the atmosphere and surface water from the equatorial Indian Ocean: Enantiomeric signatures, sources and fate. *Environmental Science and Technology,* 2013, 47, 13395−13403.
[85] Vorkamp, K; Bossi, R; Bester, K; Bollmann, UE; Boutrup, S. New priority substances of the European Water Framework Directive: Biocides; pesticides and brominated flame retardants in the aquatic environment of Denmark. *Science of the Total Environment,* 2014, 470/471, 459–468.
[86] Ying, GG; Kookana, R. Degradation of five selected endocrine–disrupting chemicals in seawater and marine sediment. *Environmental Science and Technology,* 2003, 37, 1256–1260.
[87] Basheer, C; Lee, HK; Tan, KS. Endocrine disrupting alkylphenols and bisphenol–A in coastal waters and supermarket seafood from Singapore. *Marine Pollution Bulletin,* 2004, 48, 1145–1167.
[88] Brossa, L; Marce, RM; Borrull, F; Pocurull, E. Occurrence of twenty–six endocrine–disrupting compounds in enviromental water samples from Catalonia; Spain. *Enviromental Toxicology and Chemistry,* 2005, 24, 261–267.
[89] Li, D; Dong, M; Shim, WJ; Hong, SH; Oh, JR; Yim, UH; Jeung, JH; Kanan, N; Kim, ES; Cho, SR. Seasonal and spatial distribution of nonylphenol and IBP in Saemangeum Bay, Korea. *Marine Pollution Bulletin,* 2005, 51, 966–974.
[90] Valsamaki, VI; Sakkas, VA; Albanis, TA. Determination of the pesticides considered as endocrine–disrupting compounds (EDCs) by solid phase extraction followed by gas chromatography with electron capture and mass spectrometric detection. *Journal of Separation Sciences,* 2007, 30, 1936–1946.
[91] Santiago, EC; Kwan, CS. Endocrine–disrupting phenols in selected rivers and bays in the Philippines. *Marine Pollution Bulletin,* 2007, 54, 1037–1046.

[92] Arditsoglou, A; Voutsa, D. Determination of phenolic and steroid endocrine disrupting compounds in environmental matrices. *Environmental Science and Pollution Research,* 2008, 15, 228–236.

[93] López–Darias, J; Germán–Hernández, M; Pino, V; Afonso, AM. Dispersive liquid–liquid microextraction versus single–drop microextraction for the determination of several endocrine–disrupting phenols from seawaters. *Talanta,* 2010, 80, 1611–1618.

[94] Rocha, MJ; Ribeiroand, C; Ribeiro, M. Development and optimisation of a GC–MS method for the evaluation of oestrogens and persistent pollutants in river and seawater samples. International Journal of Analytical Chemistry, 2011, 91, 1191–1205.

[95] Arditsoglou, A; Voutsa, D. Occurrence and partitioning of endocrine–disrupting compounds in the marine environment of Thermaikos Gulf, Northern Aegean Sea, Greece. *Marine Pollution Bulletin,* 2012, 64, 2443–2452.

[96] Lisboa, NS; Fahning, CS; Cotrim, G; dos Anjos, JP; de Andrade, JB; Hatje, V; da Rocha, GO. A simple and sensitive UFLC–fluorescence method for endocrine disrupters determination in marine waters. *Talanta,* 2013, 117, 168–175.

[97] Yu, L; Wu, M; Dong, W; Jin, J; Zhang, X. Reverse flow injection analysis of trace amounts of anionic surfactants in seawater without liquid–liquid extraction. *Soil and Water Research,* 2011, 6, 198–204.

[98] Place, BJ; Perkins, MJ; Sinclair, E; Barsamian, AL; Blakemore, PR; Field, JA. Trace analysis of surfactants in Corexit oil dispersant formulations and seawater. Deep–*Sea Research II*, 2014, DOI: 10.1016/j.dsr2.2014.01.015i.

[99] Beneito–Cambra, M; Ripoll–Seguer, L; Herrero–Martinez, JM; Simo–Alfonso, EF; Ramis–Ramos, G. Determination of fatty alcohol ethoxylates and alkylether sulfates by anionic exchange separation; derivatization with a cyclic anhydride and liquid chromatography. *Journal of Chromatography A,* 2011, 1218, 8511– 8518.

[100] Lara–Martin, PA; Gomez–Parra, A; Gonzalez–Mazo, E. Simultaneous extraction and determination of anionic surfactants in waters and sediments. *Journal of Chromatography A,* 2006, 1114, 205–210.

[101] Lara–Martin, PA; Gomez–Parra, A; Gonzalez–Mazo, E. Development of a method for the simultaneous analysis of anionic and non–ionic surfactants and their carboxylated metabolites in environmental samples by mixed–mode liquid chromatography–mass spectrometry. *Journal of Chromatography A,* 2006, 1137, 188–197.

[102] Lara–Martin, PA; Gomez–Parra, A; Gonzalez–Mazo, E. Pressurized liquid extraction followed by liquid chromatography–mass spectrometry for the determination of major surfactants in marine sediments. *International Journal of Environmental Analytical Chemistry,* 2005, 85, 293 – 303.

[103] González, S; Petrovic, M; Barceló, D. Simultaneous extraction and fate of linear alkylbenzene sulfonates, coconut diethanol amides, nonylphenol ethoxylates and their degradation products in wastewater treatment plants, receiving coastal waters and sediments in the Catalonian area (NE Spain). *Journal of Chromatography A,* 2004, 1052, 111–120.

[104] Li, X; Brownawell, BJ. Analysis of quaternary ammonium compounds in estuarine sediments by LC–ToF–MS: very high positive mass defects of alkylamine ions as powerful diagnostic tools for identification and structural elucidation. *Analytical Chemistry,* 2009, 81, 7926–7935.

[105] Cantarero, S; Camino–Sánchez, FJ; Zafra–Gómez, A; Ballesteros, O; Navalón, A; Vílchez, JL; Verge, C; Reis, MS; Saraiva PM. Evaluation of the presence of major anionic surfactants in marine sediments. *Marine Pollution Bulletin,* 2012, 64, 587–594.

[106] Fernández–Ramos, C; Ballesteros, O; Blanc, R; Zafra–Gómez, A; Camino–Sánchez, FJ; Navalón, A; Vílchez, JL. Determination of alcohol sulfates and alcohol ethoxysulfates in marine and river sediments using liquid chromatography–tandem mass spectrometry. *Talanta,* 2013, 115, 606–615.

[107] Traverso–Soto, JM; Lara–Martín, PA; León, VM; González–Mazo, E. Analysis of alcohol polyethoxylates and polyethylene glycols in marine sediments. *Talanta,* 2013, 110, 171–179.

[108] Montesdeoca–Esponda, S; Sosa–Ferrera, Z; Santana–Rodríguez, JJ. Combination of microwave–assisted micellar extraction with liquid chromatography tandem mass spectrometry for the determination of fluoroquinolone antibiotics in coastal marine sediments and sewage sludges samples. *Biomedical Chromatography,* 2012, 26, 33–40.

[109] Löffler, D; Ternes, TA. Determination of acidic pharmaceuticals, antibiotics and ivermectin in river sediment using liquid chromatography–tandem mass spectrometry. *Journal of Chromatography A,* 2003, 1021, 133–134.

[110] Cueva–Mestanza, R; Sosa–Ferrera, Z; Torres–Padrón, ME; Santana–Rodríguez, JJ. Preconcentration of pharmaceuticals residues in sediment samples using microwave assisted micellar extraction coupled with solid

phase extraction and their determination by HPLC–UV. *Journal of Chromatography B,* 2008, 863, 150–157.

[111] Dussault, EB; Balakrishnan, VK; Solomon, KR; Sibley, PK. Matrix effects on mass spectrometric determinations of four pharmaceuticals and personal care products in water, sediments, and biota. *Canadian Journal of Chemistry,* 2009, 87, 662–672.

[112] Prat, MD; Ramil, D; Compañó, R; Hernández–Arteseros, JA; Granados, M. Determination of flumequine and oxolinic acid in sediments and soils by microwave–assisted extraction and liquid chromatography–fluorescence. *Analytica Chimica Acta,* 2006, 567, 229–235.

[113] Morales, S; Canosa, P; Rodríguez, I; Rubi, E; Cela, R. Microwave assisted extraction followed by gas chromatography with tandem mass spectrometry for the determination of triclosan and two related chlorophenols in sludge and sediments. *Journal of Chromatography A,* 2005, 1082, 128–135.

[114] Isobe, T; Serizawa, S; Horiguchi, T; Shibata, Y; Managaki, S; Takada, H; Morita, M; Shiraishi, H. Horizontal distribution of steroid estrogens in surface sediments in Tokyo Bay. *Environmental Pollution,* 2006, 144, 632–638.

[115] Morales–Muñoz, S; Luque–Garcia, JL; Ramos, MJ; Martinez–Bueno, MJ; Luque de Castro, MD. Sequential automated focused Microwave–Assisted Soxhlet extraction of compounds with different polarity from marine sediments prior to gas chromatography mass spectrometry detection, *Chromatographia,* 2005, 62, 69–74.

[116] Nuñez, L; Tadeo, JL; Garcia–Valcarcel, AI; Turiel, E. Determination of parabens in environmental solid samples by ultrasonic–assisted extraction and liquid chromatography with triple quadrupole mass spectrometry. *Journal of Chromatography A,* 2008, 1214, 178–182.

[117] Nuñez, L; Turiel, E; Martín–Esteban, A; Tadeo, JL. Molecularly imprinted polymer for the extraction of parabens from environmental solid samples prior to their determination by high hy–ultraviolet detection. *Talanta,* 2010, 80, 1782–1788.

[118] Pintado–Herrera, MG; González–Mazo, E; Lara–Martín, PA. Determining the distribution of triclosan and methyl triclosan in estuarine settings. *Chemosphere,* 2014, 95, 478–485.

[119] Sánchez–Brunete, C; Miguel, E; Albero, B; Tadeo, JL. Analysis of salicylate and benzophenone–type UV filters in soils and sediments by simultaneous extraction cleanup and gas chromatography–mass spectrometry. *Journal of Chromatography A,* 2011, 1218, 4291– 4298.

[120] Wu, HH; Chen, HC; Ding, WH. Combining microwave–assisted extraction and liquid chromatography–ion–trap mass spectrometry for the analysis of hexabromocyclododecane diastereoisomers in marine sediments. *Journal of Chromatography A,* 2009, 1216, 7755–7760.
[121] Eiguren Fernández, A; Sosa Ferrera, Z; Santana Rodríguez, JJ. Application of microwave–assisted extraction using micellar media to the determination of polychlorinated biphenyls in marine sediments. *Analytica Chimica Acta,* 2001, 433, 237–244.
[122] Blanco, E; Casais, MC; Mejuto, MC; Cela, R. Approaches for the simultaneous extraction of tetrabromobisphenol A, tetrachlorobisphenol A, and related phenolic compounds from sewage sludge and sediment samples based on matrix solid–phase dispersion. *Analytical Chemistry,* 2006, 78, 2772– 2778.
[123] de Boer, J; Wester, PG; van der Horst, A; Leonards, PEG. Polybrominated diphenyl ethers in influents, suspended particulate matter, sediments, sewage treatment plant and effluents and biota from the Netherlands. *Environmental Pollution,* 2003, 122, 63–74.
[124] de la Cal, A; Eljarrat, E; Barceló, D. Determination of 39 polybrominated diphenylether congeners in sediment samples using fast selective pressurized liquid extraction and purification. *Journal of Chromatography A,* 2003, 1021, 165–173.
[125] King, J; Readman, JW; Zhou, JL. Dynamic behaviour of polycyclic aromatic hydrocarbons in Brighton marina, UK. *Marine Pollution Bulletin,* 2004, 48, 229–239.
[126] Huang, W; Wang, Z; Yan, W. Distribution and sources of polycyclic aromatic hydrocarbons (PAHs) in sediments from Zhanjiang Bay and Leizhou Bay, South China. *Marine Pollution Bulletin,* 2012, 64, 1962–1969.
[127] García–López, M; Rodríguez, I; Cela, R. Pressurized liquid extraction of organophosphate triesters from sediment samples using aqueous solutions. *Journal of Chromatography A,* 2009, 1216, 6986–6993.
[128] Chung, HW; Ding, WH. Determination of organophosphate flame retardants in sediments by microwave–assisted extraction and gas chromatography–mass spectrometry with electron impact and chemical ionization, *Analytical and Bioanalytical Chemistry,* 2009, 395, 2325–2334.
[129] Gevao, B; Aba, AA; Al–Ghadban, AN; Uddin, S. Depositional history of polychlorinated biphenyls in a dated sediment core from the

northwestern Arabian Gulf. *Archives of Environmental Contamination and Toxicology,* 2012, 62, 549–556.

[130] Jia, H; Sun, Y; Liu, X; Yang, M; Wang, D; Qi, H; Shen, L; Sverko, E; Reiner, EJ; Li, YF. Concentration and bioaccumulation of dechlorane compounds in coastal environment of northern China. *Environmental Science and Technology,* 2011, 45, 2613–2618.

[131] López, P; Brandsma, SA; Leonards, PEG; de Boer, J. Optimization and development of analytical methods for the determination of new brominated flame retardants and polybrominated diphenyl ethers in sediments and suspended particulate matter. *Analytical and Bioanalytical Chemistry,* 2011, 400, 871–883.

[132] Mizukawa, K; Takada, H; Takeuchi, I; Ikemoto, T; Omori, K; Tsuchiya, K. Bioconcentration and biomagnification of polybrominated diphenyl ethers (PBDEs) through lower–trophic–level coastal marine food web. *Marine Pollution Bulletin,* 2009, 58, 1217–1224.

[133] Montes, R; Rodríguez, I; Cela, R. Solid–phase microextraction with simultaneous oxidative sample treatment for the sensitive determination of tetra– to hexa–brominated diphenyl ethers in sediments. *Journal of Chromatography A*, 2010, 1217, 14–21.

[134] Oros, H; Hoover, D; Rodigari, F; Crane, D; Sericano, J. Levels and distribution of polybrominated diphenyl ethers in water, surface sediments, and bivalves from the San Francisco estuary. *Environmental Science and Technology,* 2005, 39, 33–41.

[135] Morris, CR; Allchin, BN; Zegers, JJ; Haftka H; Boon J. P; Belpaire C; Leonards PEG; van Leeuwen SPJ; de Boer J. Distribution and fate of HBCD and TBBPA brominated flame retardants in North Sea estuaries and aquatic food webs. *Environmental Science and Technology,* 2004, 38, 5497–504.

[136] Remberger, M; Sternbeck, J; Palm, A; Kaj, L; Strömberg, K; Brorströn–Lundén, E. The environmental occurrence of hexabromocyclododecane in Sweden. *Chemosphere,* 2004, 54, 9–21.

[137] Ricklund, N; Kierkegaard, A; McLachlan, MS. levels and potential sources of decabromodiphenyl ethane (DBDPE) and decabromodiphenyl ether (DecaBDE) in lake and marine sediments in Sweden. *Environmental Science and Technology,* 2010, 44, 1987–1991.

[138] Salgado–Petinal, C; García–Chao, M; Llompart, M; García–Jares, C; Cela, R. Headspace solid–phase microextraction gas chromatography tandem mass spectrometry for the determination of brominated flame

retardants in environmental solid samples. *Analytical and Bioanalytical Chemistry,* 2006, 385, 637–644.

[139] Sedôdio, P; Salomé Cabrial, M; Nogueira, JMF. Use of experimental design in the optimization of stir bar sorptive extraction for the determination of polybrominated diphenyl ethers in environmental matrices. *Journal of Chromatography A,* 2007, 1141, 259–270.

[140] Yusà, V; Pardo, O; Pastor, A; de la Guardia, M. Optimization of a microwave–assisted extraction large–volume injection and gas chromatography–ion trap mass spectrometry procedure for the determination of polybrominated diphenyl ethers, polybrominated biphenyls and polychlorinated naphthalenes in sediments. *Analytica Chimica Acta,* 2006, 557, 304–313.

[141] Zhao, Z; Zhong, G; Möller, A; Xie, Z; Sturm, R; Ebinghaus, R; Tang, J; Zhang, G. Levels and distribution of dechlorane plus in coastal sediments of the Yellow Sea, North China. *Chemosphere,* 2011, 83, 984–990.

[142] Kolpin, DW; Furlong, ET; Meyer, MT; Thurman, EM. Pharmaceuticals, hormones, and other organic wastewater contaminants in U. S. Streams, 1999–2000: A National reconnaissance. *Environmental Science and Technology,* 2002, 36, 1202–1211.

[143] Christen, V; Hickmann, S; Rechenberg, B; Fent, K. Highly active human pharmaceuticals in aquatic systems: A concept for their identification based on their mode of action. *Aquatic Toxicology,* 2010, 96, 167–181.

[144] Ahrens, L; Felizeter, S; Ebinghaus, R. Spatial distribution of polyfluoroalkyl compounds in seawater of the German Bight. *Chemosphere,* 2009, 76, 179–184.

[145] Noppe, H; Ghekiere, A; Verslycke, T; De Wulf, E; Verheyden, K; Monteyne, E; Polfliet, K; Van Caeter, P; Janssen, C; De Brabander, H. Distribution and ecotoxicity of chlorotriazines in the Scheldt Estuary (B–Nl). *Environmental Pollution,* 2007, 147, 668–676.

[146] Weigel, S; Bester, K; Hühnerfuss, H. New method for rapid solid–phase extraction of large–volume water samples and its application to non–target screening of North Sea water for organic contaminants by gas chromatography–mass spectrometry. *Journal of Chromatography A,* 2001, 912, 151–161.

[147] Wille, K; Vanden Bussche, J; Noppe, H; De Wulf, E; Van Caeter, P; Janssen, CR; De Brabander, HF; Vanhaecke, L. A validated analytical method for the determination of perfluorinated compounds in surface–, sea– and sewagewater using liquid chromatography coupled to time–of–

flight mass spectrometry. *Journal of Chromatography A,* 2010, 1217, 6616–6622.

[148] Wille, K; Noppe, H; Verheyden, K; Vanden Bussche, J; De Wulf, E; Van Caeter, P; Janssen, CR; De Brabander, H.F; Vanhaecke, L. Validation and application of an LC–MS/MS method for the simultaneous quantification of 13 pharmaceuticals in seawater. *Analytical and Bioanalytical Chemistry*, 2010, 397, 1797–1808.

[149] Torres, M; Barros, M; Campos, S; Pinto, E; Rajamani, S; Sayre, R; Colepicolo, P. Biochemical biomarkers in algae and marine pollution: A review. *Ecotoxicology and Environmental Safety,* 2008, 71, 1–15.

[150] Corcoran, J; Winter, M; Tyler, C. Pharmaceuticals in the aquatic environment: A critical review of the evidence for health effects in fish. *Critical Reviews in Toxicology,* 2010, 40, 287–304.

[151] Falandysz, J; Taniyasu, S; Gulkowska, A; Yamashita, N; Schulte–Oehlmann, U. Is fish a major source of fluorinated surfactants and repellents in humans living on the Baltic Coast? *Environmental Science and Technology* 2006, 40, 748–751.

[152] Haug, LS; Thomsen, C; Brantsaeter, AL; Kvalem, HE; Haugen, M; Becher, G; Alexander, J; Meltzer, HM; Knutsen, HK. Diet and particularly seafood are major sources of perfluorinated compounds in humans. *Environmental International,* 2010, 36, 772–778.

[153] Dussault, EB; Balakrishnan, VK; Solomon, KR; Sibley, PK. Matrix effects on mass spectrometric determinations of four pharmaceuticals and personal care products in water, sediments, and biota. *Canandian Journal of Chemistry,* 2009, 87, 662–672.

[154] McGlinchey, TA; Rafter, PA; Regan, F; McMahon, GP. A review of analytical methods for the determination of aminoglycoside and macrolide residues in food matrices, *Analytica Chimica Acta,* 2008, 624, 1–15.

[155] Bogialli, S; Di Corcia, A. Recent applications of liquid chromatography–mass spectrometry to residue analysis of antimicrobials in food of animal origin. *Analytical and Bioanalytical Chemistry,* 2009, 395, 947–966.

[156] Cañada–Cañada, F; Muñoz de la Peña, A; Espinosa–Mansilla, A. Analysis of antibiotics in fish samples. *Analytical and Bioanalytical Chemistry,* 2009, 395, 987–1008.

[157] Huerta, B; Rodríguez–Mozaz, S; Barceló, D. Pharmaceuticals in biota in the aquatic environment: analytical methods and environmental

implications. *Analytical and Bioanalytical Chemistry*, 2012, 404, 2611–2624.

[158] Ramirez, AJ; Mottaleb, MA; Brooks, BW; Chambliss, CK. Analysis of pharmaceuticals in fish using liquid chromatography–tandem mass spectrometry. *Analytical Chemistry,* 2007, 79, 3155–3163.

[159] Lahti, M; Brozinski, JM; Jylha, A; Kronberg, L; Oikari, A. Uptake from water, biotransformation, and biliary excretion of pharmaceuticals by rainbow trout. *Enviromental Toxicology and Chemistry,* 2011, 30, 1403–1411.

[160] Kallio, JM; Lahti, M; Oikari, A; Kronberg, L. Metabolites of the aquatic pollutant diclofenac in fish bile. *Environmental Science & Technology,* 2010, 44, 7213–7219.

[161] Mehinto, A; Hill, E; Tyler, C. Uptake and biological effects of environmentally relevant concentrations of the nonsteroidal anti–inflammatory pharmaceutical diclofenac in rainbow trout (*Oncorhynchus mykiss*). *Environmental Science & Technology,* 2010, 44, 2176–2182.

[162] Mimeault, C; Woodhouse, AJ; Miao, XS; Metcalfe, CD; Moon, TW; Trudeau, VL. The human lipid regulator, gemfibrozil bioconcentrates and reduces testosterone in the goldfish, *Carassius auratus*. *Aquatic Toxicology,* 2005, 73, 44–54.

[163] Owen, SF; Huggett, DB; Hutchinson, TH; Hetheridge, MJ; Kinter, LB; Ericson, JF; Sumpter, JP. Uptake of propranolol, a cardiovascular pharmaceutical, from water into fish plasma and its effects on growth and organ biometry. *Aquatic Toxicology,* 2009, 93, 217–224.

[164] Nallani, GC; Paulos, PM; Constantine, LA; Venables, BJ; Huggett, DB. Bioconcentration of ibuprofen in fathead minnow (*Pimephales promelas*) and channel catfish (*Ictalurus punctatus*. *Chemosphere,* 2011, 84, 1371–1377.

[165] Ericson, H; Thorsén, G; Kumblad, L. Physiological effects of diclofenac, ibuprofen and propranolol on Baltic Sea blue mussels. *Aquatic Toxicology,* 2010, 99, 223–231.

[166] Ramírez, AJ; Brain, RA; Usenko, S; Mottaleb, MA; O'Donnell, JG; Stahl, LL; Wathen, JB; Snyder, BD; Pitt, JL; Perez-Hurtado, P; Dobbins, LL; Brooks, BW; Chambliss, C K. Occurrence of pharmaceuticals and personal care products in fish: results of a national pilot study in the United States. *Enviromental Toxicology and Chemistry,* 2009, 28, 2587–2597.

[167] Schuetze, A; Heberer, T; Effkemann, S; Juergensen, S. Occurrence and assessment of perfluorinated chemicals in wild fish from Northern Germany. *Chemosphere,* 2010, 78, 647–652.

[168] Skarphedinsdottir, H; Gunnarsson, K; Gudmundsson, GA; Nfon, E. Bioaccumulation and biomagnification of organochlorines in a marine food web at a pristine site in Iceland, *Archives of Environmental Contamination and Toxicology*, 2010, 58, 800–809.

[169] Kirbiš, A; Marinšek, J; Flajs, V. Introduction of the HPLC method for the determination of quinolone residues in various muscle tissues. *Biomedical Chromatography,* 2005, 19, 259–265.

[170] Schultz, M; Furlong, E; Kolpin, D; Werner, S; Schoenfuss, H; Barber, L; Blazer, VS; Norris, DO; Vajda, AM. Antidepressant pharmaceuticals in two U.S. effluent–impacted streams: occurrence and fate in water and sediment, and selective uptake in fish neural tissue. *Environmental Science & Technology,* 2010, 44, 1918–1925.

[171] Nie, XP; Liu, BY; Yu, HJ; Liu, WQ; Yang, YF. Toxic effects of erythromycin, ciprofloxacin and sulfamethoxazole exposure to the antioxidant system in *Pseudokirchneriella subcapitata. Environmental Pollution*, 2013, 172, 23–32.

[172] Dussault, EB; Balakrishnan, VK; Borgmann, U; Solomon, KR; Sibley, PK. Bioaccumulation of the synthetic hormone 17α–ethinylestradiol in the benthic invertebrates *Chironomus tentans* and *Hyalella azteca. Ecotoxicology and Environmental Safety,* 2009, 72, 1635–1641.

[173] Vannini, C; Domingo, G; Marsoni, M; De Mattia, F; Labra, M; Castiglioni, S; Bracale, M. Effects of a complex mixture of therapeutic drugs on unicellular algae *Pseudokirchneriella subcapitata. Aquatic Toxicology,* 2011, 101, 459–465.

[174] Wang, M; Guo, B; Huang, Z; Duan, J; Chen, Z; Chen, B; Yao, S. Improved compatibility of liquid chromatography with electrospray tandem mass spectrometry for tracing occurrence of barbital homologous residues in animal tissues. *Journal of Chromatography A,* 2010, 1217, 2821–2831.

[175] Ramírez, AJ; Mottaleb, MA; Brooks, BW; Chambliss, CK; Analysis of pharmaceuticals in fish using liquid chromatography–tandem mass spectrometry. *Analytical Chemistry,* 2007, 79, 3155–3163.

[176] Pedersen, SN; Lindholst, C. Quantification of the xenoestrogens 4–tert–octylphenol and bisphenol A in water and in fish tissue based on microwave assisted extraction , solid–phase extraction and liquid

chromatography–mass spectrometry. *Journal of Chromatography A,* 1999, 864, 17–24.

[177] Cueva–Mestanza, R; Torres–Padrón, ME; Sosa–Ferrera, Z; Santana–Rodríguez, JJ. Microwave–assisted micellar extraction coupled with solid–phase extraction for preconcentration of pharmaceuticals in molluscs prior to determination by HPLC, *Biomedical Chromatography,* 2008, 22, 1115–1122.

[178] Fernandez–Torres, R; Lopez, MAB; Consentino, MO; Mochon, MC; Payan, MR. Enzymatic–microwave assisted extraction and high–performance liquid chromatography–mass spectrometry for the determination of selected veterinary antibiotics in fish and mussel samples. *Journal of Pharmaceutical and Biomedical Analysis,* 2011, 54, 1146–1156.

[179] Carabias–Martínez, R; Rodríguez–Gonzalo, E; Revilla–Ruiz, P; Hernández–Méndez, J. Pressurized liquid extraction in the analysis of food and biological samples. *Journal of Chromatography A,* 2005, 1089, 1–17.

[180] McEneff, G; Barron, L; Kelleher, B; Paull, B; Quinn, B. A year–long study of the spatial occurrence and relative distribution of pharmaceutical residues in sewage effluent, receiving marine waters and marine bivalves. *Science of the Total Environment,* 2014, 476–477, 317–326.

[181] Carafa, R; Wollgast, J; Canuti, E; Ligthart, J; Dueri, S; Hanke, G; Eisenreich, SJ; Viaroli, P; Zaldívar, JM. Seasonal variations of selected herbicides and related metabolites in water, sediment, seaweed and clams in the Sacca di Goro coastal lagoon (Northern Adriatic). *Chemosphere,* 2007, 69, 1625–1637.

[182] Chu, S; Metcalfe, CD. Analysis of paroxetine, fluoxetine and norfluoxetine in fish tissues using pressurized liquid extraction, mixed mode solid phase extraction cleanup and liquid chromatography–tandem mass spectrometry. *Journal of Chromatography A,* 2007, 1163, 112–118.

[183] Belfroid, A; van Velzen, M; van der Horst, B; Vethaak, D. Occurrence of bisphenol A in surface water and uptake in fish: evaluation of field measurements. *Chemosphere,* 2002, 49, 97–103.

[184] Al–Ansari, AM; Saleem, A; Kimpe, LE; Trudeau, VL; Blais, JM. The development of an optimized sample preparation for trace level detection of 17α–ethinylestradiol and estrone in whole fish tissue. *Journal of Chromatography B,* 2011, 879, 3649–3652.

[185] Subedi, B; Mottaleb, MA; Chambliss, CK; Usenko, S. Simultaneous analysis of select pharmaceuticals and personal care products in fish tissue using pressurized liquid extraction combined with silica gel cleanup. *Journal of Chromatography A,* 2011, 1218, 6278–6284.

[186] Buser, HR; Balmer, ME; Schmid, P; Kohler, M. Occurrence of UV filters 4–methylbenzylidene camphor and octocrylene in fish from various Swiss rivers with inputs from wastewater treatment plants. *Environmental Science and Technology,* 2006, 40, 1427–1431.

[187] Balmer, ME; Buser, HR; Müller, MD; Poiger, T. Occurrence of some organic UV filters in wastewater, in surface waters, and in fish from Swiss Lakes. *Environmental Science and Technology,* 2005, 39, 953–962.

[188] Wise, SA; Benner, BA; Byrd, GD; Chesler, SN; Rebbert, RE; Schantz, MM. Determination of polycyclic aromatic hydrocarbons in a coal tar standard reference material. *Analytical Chemistry*, 1988 60, 887–894.

[189] Berset, J; Kuehne, P; Shotyk, W. Concentrations and distribution of some polychlorinated biphenyls (PCBs) and polycyclic aromatic hydrocarbons (PAHs) in an ombrotrophic peat bog profile of Switzerland. *Science of the Total Environment,* 2001, 267, 67–85.

[190] Jacob, J. Method development for the determination of polycyclic aromatic hydrocarbons (PAHs) in environmental matrices. *Techniques and Instrumentation in Analytical Chemistry,* 1995, 17, 563–589.

[191] Schantz, MM; Koster, BJ; Oakley, LM; Schiller, SB; Wise, SA. Certification of polychlorinated biphenyl congeners and chlorinated pesticides in a whale blubber standard reference material. *Analytical Chemistry,* 1995, 67, 901–910.

[192] Fuoco, R; Ceccarini, A; Onor, M; Lottici, S. Supercritical fluid extraction combined on–line with cold–trap gas chromatography/mass spectrometry. *Analytica Chimica Acta,* 1997, 346, 81–86.

[193] Fuoco, R; Ceccarini, A; Onor, M; Marrara, L. Analysis of priority pollutants in environmental samples by on–line supercritical fluid chromatography cleanup–cryo–trap–gas chromatography–mass spectrometry. *Journal of Chromatography A,* 1999, 846, 387–393.

[194] Wang, S; Oakes, KD; Bragg, LM; Pawliszyn, J; Dixon, G; Servos, MR. Validation and use of in vivo solid phase micro–extraction (SPME) for the detection of emerging contaminants in fish. *Chemosphere,* 2011, 85, 1472–1480.

[195] Ningsun Zhou, S; Oakes, KD; Servos, MR; Pawliszyn, J. Application of solid–phase microextraction for in vivo laboratory and field sampling of

pharmaceuticals in fish. *Environmental Science and Technology,* 2008, 42, 6073–6079.

[196] Quinete, N; Wu, Q; Zhang, T; Yun, SH; Moreira, I; Kannan, K. Specific profiles of perfluorinated compounds in surface and drinking waters and accumulation in mussels, fish, and dolphins from southeastern Brazil. *Chemosphere,* 2009, 77, 863–869.

[197] So, MK; Taniyasu, S; Lam, PKS; Zheng, GJ; Giesy, JP; Yamashita, N. Alkaline digestion and solid phase extraction method for perfluorinated compounds in mussels and oysters from South China and Japan. *Archives of Environmental Contamination and Toxicology,* 2006, 50, 240–248.

[198] Stancheva, M; Georgieva, S; Makedonski, L. Persistent organic pollutants–PCBs and DDTs in fish from Danube river and from black sea, Bulgaria, in: CBU International Conference on Integration and Innovation in Science and Education, 2013, 354–361.

[199] Teijon, G; Candela, L; Tamoh, K; Molina–Díaz, A; Fernández–Alba, AR. Occurrence of emerging contaminants, priority substances (2008/105/CE) and heavy metals in treated wastewater and groundwater at Depurbaix facility (Barcelona, Spain). *Science of the Total Environment*, 2010, 408, 3584–3595.

[200] Pastor–Navarro, N; Maquieira, A; Puchades, R. Review on immunoanalytical determination of tetracycline and sulfonamide residues in edible products. *Analytical and Bioanalytical Chemistry,* 2009, 395, 907–920.

[201] Huet, AC; Delahaut, P; Fodey, T; Haughey, SA; Elliott, C; Weigel, S. Advances in biosensor–based analysis for antimicrobial residues in foods. *TrAC Trends Analytical Chemistry,* 2010, 29, 1281–1294.

[202] Kantiani, L; Farré, M; Barceló, D. Analytical methodologies for the detection of β–lactam antibiotics in milk and feed samples. *TrAC Trends Analytical Chemistry,* 2009, 28, 729–744.

[203] Cháfer–Pericás, C; Maquieira, A; Puchades, R; Miralles, J; Moreno, A; Pastor–Navarro, N; Espinós, F. Immunochemical determination of oxytetracycline in fish: comparison between enzymatic and time–resolved fluorometric assays. *Analytica Chimica Acta,* 2010, 662, 177–185.

[204] Cháfer–Pericás, C; Maquieira, Á; Puchades, R; Miralles, J; Moreno, A. Fast screening immunoassay of sulfonamides in commercial fish samples. *Analytical and Bioanalytical Chemistry,* 2010, 396, 911–921.

[205] Huet, AC; Charlier, C; Singh, G; Godefroy, SB; Leivo, J; Vehniäinen, M; Nielen, MW; Weigel, S; Delahaut, P. Development of an optical surface plasmon resonance biosensor assay for (fluoro)quinolones in egg, fish, and poultry meat. *Analytica Chimica Acta,* 2008, 623, 195–203.

[206] Delépée, R; Pouliquen, H; Le Bris, H. The bryophyte *Fontinalis antipyretica Hedw.* bioaccumulates oxytetracycline, flumequine and oxolinic acid in the freshwater environment. *Science of the Total Environment,* 2004, 322, 243–253.

[207] Rambla–Alegre, M; Peris–Vicente, J; Esteve–Romero, J; Carda–Broch, S. Analysis of selected veterinary antibiotics in fish by micellar liquid chromatography with fluorescence detection and validation in accordance with regulation 2002/657/EC. Food Chemisty, 2010, 123, 1294–1302.

[208] Schröder, U; Machetzki, A. Determination of flumequine, nalidixic acid and oxolinic acid in shrimps by high–performance liquid chromatography with fluorescence detection. *European Food Research and Technology,* 2007, 225, 627–635.

[209] Kirbiš, A; Marinšek, J; Cerkvenik, V. Introduction of the HPLC method for the determination of quinolone residues in various muscle tissues. *Biomedical Chromatography,* 2005, 19, 259–265.

[210] Richardson, S; Ternes, TA. Water analysis: emerging contaminants and current issues, Analytical Chemistry, 2014, 83, 4614–4648.

[211] Kümmerer, K. Antibiotics in the aquatic environment–a review–part I. *Chemosphere,* 2009, 75, 417–434.

[212] Hao, C; Zhao, X; Yang, P. GC–MS and HPLC–MS analysis of bioactive pharmaceuticals and personal–care products in environmental matrices. *Trends Analytical Chemistry,* 2007, 26, 569–580.

[213] Contardo–Jara, V; Lorenz, C; Pflugmacher, S; Nützmann, G; Kloas, W; Wiegand, C. Exposure to human pharmaceuticals Carbamazepine, Ibuprofen and Bezafibrate causes molecular effects in *Dreissena polymorpha. Aquatic Toxicology,* 2011, 105, 428–437.

[214] Vernouillet, G; Eullaffroy, P; Lajeunesse, A; Blaise, C; Gagné, F; Juneau, P. Toxic effects and bioaccumulation of carbamazepine evaluated by biomarkers measured in organisms of different trophic levels. *Chemosphere,* 2010, 80, 1062–1068.

[215] Brooks, BW; Chambliss, CK; Stanley, JK; Ramírez, A; Banks, KE; Johnson, RD; Lewis RJ. Determination of select antidepressants in fish

from an effluent–dominated stream. *Enviromental Toxicology and Chemistry,* 2005,24, 464–469.

[216] Bringolf, RB; Heltsley, RM; Newton, TJ; Eads, CB; Fraley, SJ; Shea, D; Gregory Cope, W. Environmental occurrence and reproductive effects of the pharmaceutical fluoxetine in native freshwater mussels. *Enviromental Toxicology and Chemistry,* 2010, 29, 1311–1318.

[217] Lahti, M; Brozinsky, JM; Jylhä, A; Kronberg, L; Oikari, A. Uptake from water, biotransformation, and biliary excretion of pharmaceuticals by rainbow trout. *Enviromental Toxicology and Chemistry,* 2011, 30, 1403–1411.

[218] Lajeunesse, A; Vernouillet, G; Eullaffroy, P; Gagnon, C; Juneaub, P; Sauvé, S. Determination of carbamazepine in aquatic organisms by liquid–liquid extraction and liquid chromatography–tandem mass spectrometry. *Journal of Environmental Monitoring,* 2009, 11, 723–725.

[219] Kaufmann, A; Butcher, P; Maden, K. Determination of aminoglycoside residues by liquid chromatography and tandem mass spectrometry in a variety of matrices. *Analytica Chimica Acta,* 2012, 711, 46–53.

[220] Mehinto, A; Hill, E; Tyler, C. Uptake and biological effects of environmentally relevant concentrations of the nonsteroidal anti–inflammatory pharmaceutical diclofenac in rainbow trout (*Oncorhynchus mykiss*). *Environmental Science & Technology,* 2010, 44, 2176–2182.

[221] Fuoco, R; Giannarelli, S; Onor, M; Ceccarini, A; Carli, V. Optimized cleanup methods of organic extracts for the determination of organic pollutants in biological samples, *Microchemical Journal,* 2005, 79, 69–76.

[222] Celiz, M; Tso, J; Aga, D. Pharmaceutical metabolites in the environment: Analytical challenges and ecological risks. *Enviromental Toxicology and Chemistry,* 2009, 28, 2473–2484.

[223] Dussault, E; Balakrishnan, V; Solomon, K; Sibley, P. Matrix effects on mass spectrometric determinations of four pharmaceuticals and personal care products in water, sediments, and biota. *Canadian Journal of Chemistry,* 2009, 87, 662–672.

[224] Gu, M; Mitchell, R; Kim, B. Whole–cell–based biosensors for environmental biomonitoring and application. *Advances in Biochemical Engineering, Biotechnology,* 2004, 87, 269–305.

[225] Scognamiglio, V; Pezzotti, I; Pezzotti, G; Cano, J; Manfredonia, I; Bounasera, K; Arduini, F; Moscone, D; Palleschi, G; Giardi, MT. Towards an integrated biosensor array for simultaneous and rapid multi–

analysis of endocrine disrupting chemicals. *Analytica Chimica Acta,* 2013, 751, 161–170.

[226] Cabaj, J; Soloducho, J. Layered biosensor construction, in: T. Rinken (Ed.), State of the art in biosensors–general aspects, Rijeka, Croatia, 2013, 37–66.

[227] Poynton, H; Vulpe, C. Ecotoxicogenomics: emerging technologies for emerging contaminants. *Journal of the American Water Resources Association,* 2009, 45, 83–96.

[228] McAllister, L; Semple, K. Measurement of bioaccessibility of organic pollutants in soil, *Methods in Molecular Biology,* 2010, 599, 1–14.

[229] Rayu, S; Karpouzas, DG; Singh, BK. Emerging technologies in bioremediation: constraints and opportunities. *Biodegradation,* 2012, 23, 917–926.

[230] Perelo, LW. Review: In situ and bioremediation of organic pollutants in aquatic sediments. *Journal of Hazardous Materials,* 2010, 177, 81–89.

[231] Lovley, D. Anaerobic benzene degradation. *Biodegradation,* 2000, 11, 107–116.

[232] Jeong, S. Evaluation of the use of capillary numbers for quantifying the removal of DNAPL trapped in a porous medium by surfactant and surfactant foam floods. *Journal of Colloid and Interface Science,* 2005, 282, 182–187.

[233] Johnson, D; Pedit, J; Miller, C. Efficient near–complete removal of DNAPL from three–dimensional, heterogeneous porous media using a novel combination of treatment technologies. *Environmental Science and Technology,* 2004, 38, 5149–5156.

[234] Nakajima, F; Baun, A; Ledin, A; Mikkelsen, P. A novel method for evaluating bioavailability of polycyclic aromatic hydrocarbons in sediments of an urban stream. *Water Science and Technology,* 2005, 51, 275–281.

[235] Ahrens, M; Hertz, J; Lamoureux, E; Lopez, G; McElroy, A; Brownawell, B. The role of digestive surfactants in determining bioavailability of sediment–bound hydrophobic organic contaminants to 2 deposit–feeding polychaetes. *Marine Ecology Progress Series,* 2001, 212, 145–157.

[236] Zhang, Y; Miller, R. Effect of rhamnolipid (biosurfactant) structure on solubilization and biodegradation of n–alkanes. *Environmental Microbiology,* 1995, 61, 2247–2251.

[237] Wong, J; Fang, M; Zhao, A; Xing, B. Effect of surfactants on solubilization and degradation of phenanthrene under thermophilic conditions. *Journal of Environmental Quality,* 2004, 33, 2015–2025.

[238] Yu, H; Zhu, L; Zhou, W. Enhanced desorption and biodegradation of phenanthrene in soil–water systems with the presence of anionic–nonionic mixed surfactants. *Journal of Hazardous Materials,* 2007, 142, 354–361.

[239] Aronstein, B; Calvillo, Y; Alexander, M. Effect of surfactants at low concentrations on the desorption and biodegradation of sorbed aromatic compounds in soil. *Environmental Science and Technology,* 1991, 25, 1728–1731.

[240] Willumsen, P; Karlson, U; Pritchard, P. Response of fluoranthene–degrading bacteria to surfactants. *Applied Microbiology and Biotechnol,* 1998, 50, 475–483.

[241] Zheng, Z; Obbard, J. Evaluation of an elevated non–ionic surfactant critical micelle concentration in a soil/aqueous system. *Water Research,* 2002, 36, 2667–2672.

[242] Major, D; McMaster, M; Cox, E; Edwards, E; Dworatzek, S; Hendrickson, E; Star, MG; Payne, JA; Buonamici, LW. Field demonstration of successful bioaugmentation to achieve dechlorination of tetrachloroethene to ethene. *Environmental Science and Technology,* 2002, 36, 5106–5116.

[243] Juhasz, A; Naidu, R. Bioremediation of high molecular weight polycyclic aromatic hydrocarbons: a review of the microbial degradation of benzo[a]pyrene. *International Biodeterioration & Biodegradation,* 2000, 45, 57–88.

[244] Tam, N; Wong, Y. Effectiveness of bacterial inoculum and mangrove plants on remediation of sediment contaminated with polycyclic aromatic hydrocarbons. *Marine Pollution Bulletin,* 2008, 57, 716–726.

[245] Thompson, I; Van der Gast, C; Ciric, L; Singer, A. Bioaugmentation for bioremediation: the challenge of strain selection. *Environmental Microbiology,* 2005, 7, 909–915.

[246] Van der Gast, C; Whiteley, A; Strakey, M; Knowles, C; Thompson, I. Bioaugmentation strategies for remediating mixed chemical effluents. Biotechnol Progress, 2003, 19, 1156–1161.

[247] Van der Gast, C; Whiteley, A; Thompson, I. Temporal dynamics and degradation activity of an bacterial inoculum for treating waste metal fluid. *Environmental Microbiology,* 2004, 6, 254–263.

[248] Fulthorpe, R; Wyndham, R. Transfer and expression of the catabolic plasmid pbrc60 in wild bacterial recipients in a fresh–water ecosystem. *Applied and Environmental Microbiolgy,* 1991, 57, 1546–1553.

[249] Brokamp, A; Schmidt, F. Survival of alcaligenes xylosoxicans degrading 2,2– dichloropropionate and horizontal transfer of its halidohydrolase gene in a soil microcosm. *Current Microbiology,* 1991, 22, 299–306.

[250] Meagher, R. Phytoremediation of toxic elemental and organic pollutants. *Current Opinion in Plant Biology,* 2000, 3, 153–162.

[251] Anderson, T; Walton, B. Comparative fate of [14C] trichoroethylene in the root zone of plants from a former solvent disposal site. *Enviromental Toxicology and Chemistry,* 1996, 12, 2041–2047.

[252] Nzengung, V; Wolfe, L; Rennels, D; McCutcheon, S; Wang, C. Use of aquatic plants and algae for decontamination of waters polluted with chlorinated alkanes. *International Journal of Phytoremediation,* 1999, 1, 203–226.

[253] Aken, B; Transgenic plants for phytoremediation: helping nature to clean-up environmental pollution. *Trends in Biotechnology,* 2008, 26, 225–227.

[254] Yamamoto, T; Goto, I; Kawaguchi, O; Minagawa, K; Ariyoshi, E; Matsuda, O. Phytoremediation of shallow organically enriched marine sediments using benthic microalgae. *Marine Pollution Bulletin,* 2008, 57, 108–11.

In: Marine Pollution
Editor: Dominic E. Jefferson
ISBN: 978-1-63321-520-7

Chapter 4

OIL SPILLS: BEHAVIOR OF OIL, IMPACT, DETECTION, TRACKING AND MANAGEMENT

D. Kitsiou [*] ***and M. Karydis***
Department of Marine Sciences, School of the Environment, University of the Aegean, Mytilene, Lesvos, Greece

ABSTRACT

Oil related activities are potential threats for causing oil spills with serious environmental, economic and social impact. Oil spill pollution has been an issue of concern to coastal states as well as to international organizations and therefore work on prevention, detection, toxicity monitoring and impact assessment have been implemented at both national and international level. The contribution of new technologies, especially Geographical Information Systems and remote sensing, has improved both tracking and management operations. In the present work, oil behavior, monitoring and pollution impact of the marine environment have been reviewed. The importance of applying remote sensing and mapping techniques, modeling and GIS tools to either forecast oil trajectories or assess the vulnerability of coastal areas to oil and the impact on the marine ecosystem are presented. The advantages and shortcomings of different methodologies are discussed; however, the need for the implementation of integrated approaches seems the most effective approach. The advantage of applying such approaches in the

[*] E-mail: dkit@aegean.gr.

framework of a GIS system and especially a webGIS system are discussed, since direct communication and interaction among authorities, stakeholders and the wider public is essential in the case of an emergency situation.

Keywords: Oil pollution, Oil spill detection, GIS, Remote sensing, Modeling, Risk management, Environmental quality

1. INTRODUCTION

Petroleum has become a vital commodity since the beginning of the 20th century. The motorization of the western communities brought a revolution to transportation (tracks, cars, ships, trains and planes) and industrial development; it also formed the basis for production of fertilizers, plastics and medicines. It is also a source of energy for electricity production and house heating. Oil production showed an exponential trend of increase especially during the second half of the 20th century to satisfy the demand (NRC, 2003). However, oil production and consumption do not take place in the same area: the main consumers are the heavily industrialized countries that are the United States, Western European Countries and Japan. On the other hand the main oil fields are located in the Middle East, South America (mainly Venezuela) and Russia. Oil production approached 2.4 billion tons in 2005 and 60% of the oil consumed in the world was transported by tankers (Burgherr, 2007). The releases from the transportation of petroleum nowadays account for 12% of the total oil inputs into the sea (NRC, 2003). Oil marine pollution forms an environmental threat for the marine environment due to the number and size of vessels involved in oil transportation. The average size of oil tankers by the end of the 19th century was about 3,000 tones dwt, at the beginning of the 20th century about 10,000 dwt, in 30's, 12,000 and in 70's exceeded 300,000 tones dwt (Nelson-Smith, 1972). Nowadays there are about 3,500 tankers available on the international transport market accounting for an overall world tanker fleet capacity about 280 million dwt tons. The VLCC's (Very Large Crude Carriers) account for about 30% of the oil being carried. VLCC's (about 2 million barrels per ship) are mainly used for oil transportation from the Middle East to Europe and Pacific Asia. Smaller tankers are used from Latin America (Venezuela and Mexico) to the United States. The average journey from the Middle East to Pacific Asia takes about three weeks whereas journeys in the Atlantic Basin about two weeks.

The number of vessels carrying oil, their size as well as the marine traffic especially in straits and canals increase the possibility of an accident. Two of the world oil transport points known as the “choke points” are located in the Mediterranean: the Straits of Istanbul and the Suez Canal (EEA, 1999); the first connects the Black Sea with the Mediterranean and the second connects the Red Sea with the Mediterranean. Apart from the two choke points, it has been estimated that 220,000 vessels more than 200 tonnes dwt cross annually the Mediterranean accounting for 20% of the oil pollution in the area. It has been estimated that 250-300 tankers daily cross the Mediterranean. As a consequence 15 maritime accidents are recorded annually involving ships causing oil pollution. The oil related accidents have increased since the mid 1960’s for a number of reasons: (a) increase of maritime transport in general (b) increase of oil production and transportation (c) increase of the size of tankers and (d) improved surveillance and reporting coverage. The number of 737 oil spills reported between 1970-2004, caused by boats of at least 700 tonnes dwt was much bigger than the four oil spills reported between 1960 and 1964 (Burgherr, 2007).

In addition to oil transportation by tankers, offshore platforms are also a potential danger. It has been estimated (NRC, 2003) that about 8,300 fixed and floating platforms are operating worldwide. Historically the two more serious accidental oil spills have happened in the Gulf of Mexico: the Intox-1 well blowout in 1979 followed by a release of about 480,000 tonnes over a 10 month period; the Deepwater Horizon oil spill (also known as the BP oil spill) is the largest accidental oil spill in the history of petroleum industry and resulted from the explosion of Deepwater Horizon on April 20, 2010. The leak amounted to about 780,000 m^3. Nevertheless accidental oil spills from platforms account only for the 3% of petroleum inputs worldwide (NRC, 2003).

The impact caused by oil spills is both ecological and economic. The direct ecological effects are mainly on birds, marine mammals, fish and invertebrates. Long term and short term effects of petroleum on marine organisms have been studied since a long time ago both in the field (Butler1978; Mearns et al., 2010) as well as in laboratory cultures (Karydis, 1980; Karydis, 1981). Experiments in enclosures (Lacaze, 1969; 1974; 1983) showed relative resistance of phytoplankton to the oil and at the same time serious impacts on zooplankton especially copepods. Similar work (Dahl et al., 1983) showed inhibition on the growth of diatoms and copepods. Toxicity testing of hydrocarbons on zooplankton communities (Hook et al., 1986) indicated oil treatment effects on individual species but not remarkable effects on species richness and diversity measures. Extensive field work has also followed serious accidents such as the *Sea Empress* tanker accident in the Irish Sea (1996) followed by a release of 72,000 tonnes of crude oil into the marine environment most of which was not released (Batten et al., 1998) and the *Exxon Valdez* oil spill in Alaska (1989) where 42,000 tonnes of crude oil were spilled (Monson et al., 2000).

The direct economic impact of oil pollution focuses on fishing, aquaculture and tourism. In addition to the environmental and economic impacts, the social impacts are equally severe: oil spills induce what is known as "social amplification". This phenomenon concerns environmental stakeholders, even the general public and can be intensified by social values and furthermore by the politics of media coverage (Anderson, 2002; Leschine, 2002; Kirby and Law, 2010).

As oil spills have been of concern to coastal states and international organizations the International Convention for Prevention of Pollution from ships (MARPOL 73/78) was adopted; the increase of the amounts of oil transported by sea and the Torrey Canyon disaster in 1967 resulting into an oil spill of 120,000 tonnes played an important role for the adoption of the convention (Holland and Pugh, 2010). MARPOL sets specifications on the design and construction of commercial vessels (more than 100 gross tonnes) and regulations on the operation of the vessels. In 1990 was estimated that these regulations had decreased the load of spilled oil by a percentage ranging between 80 and 99 percent (NRC, 2003). Over the last few decades a number of pollution control agencies have used ships and aircraft observations for detection of oil spills. However, there are shortcomings concerning both types of observations: limited coverage and high operational cost (Ohowa, 2009). Satellite surveillance methods proved to be more effective as they provide wide spectral coverage, they do not depend on weather conditions and they can regularly revisit the area (Ohowa, 2009). Satellite data in combination with GIS techniques and modeling have been applied to assess and monitor oil spills in the marine and coastal environments, revealing the need for integrated approaches and the development of operational systems (Zodiatis et al., 2012).

The objectives of this chapter are to (a) review the impact of oil pollution on marine organisms, (b) review the methodologies applied for monitoring and assessment of oil spills and forecasting of oil spill trajectories (c) reveal the importance of GIS and in particular of webGIS as a framework for the application and integration of the above methodologies.

2. Behavior of Oil at Sea

Oil is a complex mixture of organic compounds. Hydrocarbons form 98 percent of the crude oil content. There are also hydrocarbon derivatives containing oxygen, sulfur, nitrogen of combination of these compounds (Nelson-Smith, 1972). When crude oil is spilled into the marine environment, it spreads, transported away and eventually weathers that is it undergoes through a number of changes as a result of physical chemical and biological processes (Berridge et al., 1968; Blumer et al. 1973; Davis and Gibbs, 1975). This means that the area occupied by an oil spill, the thickness and chemical composition of the spill change rapidly with time. Physical processes affecting weathering of the crude oil are evaporation, emulsification and dissolution; whereas chemical processes include oxidation, particularly photo-oxidation and biological processes mainly focus on bacterial degradation and hydrocarbon uptake (Table 1).

Table 1. Processes affecting the behavior and fate of oil at sea

Process	Description of the process	References
Spreading	Horizontal dispersion dominated by gravity, viscous and surface tension forces	Berridge et al. (1968) Fay (1969)
Evaporation	Loss of volume during the first few days after the accident usually ranging between 40 and 70 percent of the initial volume	Fingas (1995)
Emulsification	Formation of water-in-oil emulsion mainly in the form of droplets suspended in the sea water	McKee (1956) Berridge et al. (1968) Karydis (1982a) Fingas et al. (2000)
Photo-oxidation	Oxidation of carbon atoms in hydrocarbon molecules on the sea surface catalyzed by solar radiation	Karydis (1982b) Lacaze and Villedon de NeVde (1976)
Bio-degradation	Breaking down of hydrocarbon molecules primarily by bacteria and fungi, either for energy production of detoxification	Leahy and Colwell (1990) Atlas and Bartha (1992) Prince (1993) Rand (1995)
Dissolution	"Dissolution is the chemical stabilization of oil components in water" (NRC, 2003)	Brookman et al. (1985) Hibbs et al. (1999)

Spreading is the dispersion of oil initially along the sea surface. Spreading can be divided into three different phases (Fay, 1969): the inertial phase dominated by gravity forces, the viscous phase dominated by both gravity and viscous forces and the surface tension dominated by surface tension spreading. Spreading is a continuous process and in theory an oil slick can be expected to spread until a monomolecular layer is formed (Beridge et al., 1968). Once the oil has spread to a critical thickness which is about 8mm, surface tension becomes the main factor promoting spread. Wave action breaks the oil slicks into smaller patches at sea and therefore spreading is a self-retarding process. Experiments by the American Petroleum Institute (API, 1963), have shown that in calm water, the spilled oil will eventually film out to a thickness of $1 x 10^{-4}$ cm and at this thickness the oil disappears from the surface within 24 hours. Temperature, wind velocity and wave action are the principal factors affecting spreading. There are nowadays numerous models describing the spreading of an oil slick (NRC, 2003); some of them include processes that require spreading rates of the oil slicks (Plutchak and Kolpak, 1981).

Evaporation is the most important process as far as the mass balance is concerned. It has been estimated that during the first few days following the accident light crudes can lose as much as 75 percent of their initial volume, medium crudes about 40 percent, but heavy or residual oils lose no more than 10 percent of their initial volume. As crude oil is a mixture of hundred compounds that vary from source to source, it is difficult to develop predictive models for oil spill evaporation (Fingas, 1995).

Mechanical agitation leads to the formation of oil-in-water emulsion where oil at dispersed phase becomes suspended in the form of oil droplets in the water phase. The average droplet diameter in such an oil-in-water emulsion is about 0.5 μm, its volume $6x10^{-14}$ ml and its surface area $8x10^{-9}$ cm^2 (McKee, 1956). Crude oils contain various minor constituents and impurities that function as emulsifying agents (Nelson-Smith, 1972). It has also been found that phytoplankton (especially diatoms), play an important role in the formation and stability of oil-in-water emulsions (Karydis, 1982a). Oil droplets are hydrated and if their water content varies between 30-50 percent, the droplets flow freely but if their water content is as high as 60 to 85 percent, it forms grease-like material often with a paler color, known among the spill workers as "chocolate mousse" or simply "mousse" (NRC, 2003). Asphaltenes tend to form this type of emulsion and is characterized as a rather stable form (Fingas et al., 2000). Oil disperses through the emulsification process in the water column leaving the water surface. It has been estimated that 5 tons of oil may be dispersed into the top 5m layer beneath a typical 100 ton oil slick during the first day (Dodd, 1971).

It is known that both paraffins and aromatic hydrocarbons with suitable side chains are subject to atmospheric oxidation. However, most of the oxidation of hydrocarbons takes place through photo-oxidation by sunlight. Photo-oxidation is not a specific reaction but a family of light-catalyzed reactions oxidizing the reduced carbon of petroleum hydrocarbons (NRC, 2003). Photo-oxidation requires solar radiation and light absorbing molecules (chromophores). Some heavy metals present in the oil such as vanadium, act as catalysts though other ingredients such as sulfur, are inhibitors. The effective wavelengths vary between 300 and 350 nm (Parker et al., 1971). It has been estimated that a 100 ton oil slick spreading to an area of 8 km^2 at an average thickness of 0.02 mm, loses about one ton per day through photo-oxidation. In spite of the low percentage of the photo-oxidized oil, the environmental impact is rather serious. Photo-chemically oxidized hydrocarbons produce polar ketones, aldehydes, carboxyl acids and esters. These compounds are water soluble and they disperse in the water column becoming available to marine organisms. There is strong experimental evidence that photo-oxidized products are far more toxic than the parent hydrocarbons (Lacaze and Villedon de NeVde, 1976; Karydis 1982b). Hydrocarbon photo-oxidation processes are complex but they have been summarized a long time ago (NRC, 1985).

It has been pointed out since 60s (Pilpel, 1968) that although millions of tons of oil were spilt in the marine environment since the beginning of the 20th century, there is little evidence that oil is accumulated anywhere at big quantities; physical processes can only induce oil transportation. The answer was given by Parker et al. (1971), who found out that two biological processes were involved in the oil removal from the water column: hydrocarbon bio-degradation and hydrocarbon uptake by the marine organisms. There are two different mechanisms concerning biological oxidation of hydrocarbons (a) biological oxidation for energy production. Micro-organisms, mainly bacteria and fungi, and to a lesser extent phytoplankton are involved in these processes and (b) The breaking down of hydrocarbons for detoxification processes: the metabolic products are water soluble and therefore can be easily excreted by the organism. Larger organisms such as mollusks and fish have enzymes efficient to break down PAHs. These enzymic systems can metabolize PAHs into water soluble compounds and excrete them into the marine environment (Buhler and Williams, 1988; Rand, 1995). The metabolic footpaths of hydrocarbon biodegradation have been reviewed by Leahy and Colwell (1990) as well as by Atlas and Bartha (1992). Bacteria degrade any hydrocarbons from methane to the heavier residues. However, the higher the molecular weight of hydrocarbons, the more difficult for the bacteria to decompose the molecular structure. Under field conditions rates of degradation have been reported to vary between 0.001 and 60g per m^3 per day (Atlas and Bartha, 1992). Paraffins are more easily degradable but aromatics can also be decomposed. Biodegradation rates of hydrocarbons by micro-organisms in the marine environment depend not only on environmental factors but also on prior exposure of the organisms to oil spills (such as areas near harbors, refineries and oil terminals). Prior exposure decreases the lag phase of the organisms to biodegradation as opposed to micro-organisms in pristine environments where larger lags have been observed (NRC, 2003). Among environmental factors (light, temperature, nutrient concentrations, salinity, pressure), temperature seems to play a principal role; at low temperatures the increase in oil viscosity and decrease in solubility limit the bio-availability of the oil. In addition, low rates of evaporation that is presence of toxic volatile compounds at relatively high concentrations can have toxic effects on bio-degrading micro-organisms. Low temperatures also slow down enzymic activities and therefore the rates of reactions bio-degrading hydrocarbons.

Hydrocarbon uptake is another path for hydrocarbon removal from the water column. It was observed that oil uptake by zooplanktonic organisms after the "Torrey Canyon" accident accounted for 0.1-1.0 percent of the dry weight of the organisms (Parker et al., 1971). It has been estimated that copepods, a dominant group of zooplanktonic organisms can remove oil at a rate of 3 tons.km^{-2}.day. Hydrocarbon uptake by organisms has been confirmed both in field studies and laboratory experiments (Karydis, 1980).

A small fraction of crude oil can be present in the water column as stable component. Although a small fraction of the spilled oil is soluble, aromatic hydrocarbons can be particularly dangerous due to their high toxicity. The solubility of crude oils depends on the type of crude, temperature, the molecular structure of each compound and the molecular weight (Brookman et al., 1985). Dissolution in the form of oil droplets and soluble fractions is useful information as it is incorporated in modeling oil slick behavior (Hibbs et al., 1999).

3. Impact of Oil Pollution on Marine Organisms

Information on the effects of oil pollution on marine organisms has been obtained from both field studies and laboratory experiments (Table 2). Field work i.e. effects of oil on incidentally or chronically polluted habitats, provides a more realistic view on the effects at ecosystem level. However, ecosystem complexity and variability complicate the understanding of pollution and toxicity mechanisms. On the other hand toxicity tests using laboratory cultures are based on simplified experimental designs focusing on recording deaths or physiological effects such as photosynthesis, respiration of enzymic inhibition (Karydis, 1978). In addition, the choice of organisms is usually depended on laboratory routines and on how easily the test organisms can be kept under laboratory conditions; it is therefore obvious that test species are not necessarily the most important from the ecological point of view. In addition, large species or species with long life cycles are not suitable for laboratory work. A wide diversification of experimental conditions increases the degree of difficulty when results from different experiments have to be compared. Temperature, type of media, type of container and a variety of methods to prepare aqueous media are the main sources of experimental variation. It has been documented from the early stages of experimentation on oil toxicity that dispersion of oils in aqueous media using emulsifiers is an additional confusing element as the dispersant is sometimes more toxic than the oil itself (Nelson-Smith, 1972; Mommaerts-Billiet, 1973). Different techniques both physical (fractional media approach, low energy media, high energy media) and chemical methods (preparation of water soluble fractions using emulsifiers) for preparing aqueous media for oil toxicity testing and the influence of the method to the composition of the water soluble fractions have been described by Girling (1989). Oil dispersants nowadays are characterized by low toxicity but nevertheless they increase the bioavailability of hydrocarbons (Schein et al., 2009).

Phytoplankton seems to be rather resistant to oil. Early work on the effects of oil on primary productivity in the Santa Barbara Channel (California) showed shortly after the accident normal pattern of phytoplankton distribution according to the distribution of nutrients (Ogouri and Kanter, 1971). It has also been reported that some phytoplankton species (*Asterionella, Cyclotella, Fragilaria, Meridion, Navicula and Euglena*) were thriving in areas with high levels of oil pollution (McCauley, 1966). By contrast flagellates seem to be far more sensitive. Soto et al. (1975) have reported loss of motility of flagellates after exposure to oil extracts and naphthalene. Morphological abnormalities of diatoms and flagellates grown in cultures after exposure to oil spills (Smith, 1968) or in presence of oil have been reported (Karydis and Fogg, 1980). A 45-50% decrease in chlorophyll concentrations lasting for two weeks followed the *Hebei Spirit* oil spill in South Korea (Lee et al., 2009). Oil pollution experiments have also been carried out in mesocosms systems for a better understanding of ecosystemic processes under the stress of hydrocarbons (Lacaze, 1983). The results showed that although microalgae and ciliates were highly resistant to oil, copepods (zooplankton) were rather sensitive. Testing the Ecofisk crude oil in enclosed planktonic ecosystems (Dahl et al., 1983) also showed inhibition in the growth of copepods. As general trend, higher sensitivity in heterotrophs rather than in autotrophs has been reported (Batten et al., 1998).

Table 2. Oil toxicity on marine organisms

Group of organisms	Response	References
Phytoplankton	Photosynthetic inhibition	Gonzalez-Meler et al. (2009)
	Effects on respiration	Karydis (1978)
	Effects on primary productivity	Ogouri & Kanter (1971)
	Loss of motility of flagellates	Soto et al. (1975)
Zooplankton	Toxic effects	Atlas and Bartha (1987)
	Effects on growth of copepods	Dahl et al. (1983)
Zoobenthos	Reduction in scope of growth	Donkin et al. (1990)
	Reduction in recruitment	Krebs & Burns (1977)
	Tumors – pathological disorders	Meador et al. (1995)
Phytobenthos	Impact on macroalgal communities	Diez et al. (2009)
Fishes	Mortality of fish embryos	Heintz et al. (1999)
	Delayed growth	Heintz et al. (1999)
	Toxopathic effects on estuarine fishes	Johnson et al. (2001)
Sea Birds	Seabird mortality	Burger (1993)
	Reduced reproductive success	Fry & Addiego (1987)
Marine Mammals	Effects on penguins	Westphal & Rowan (1970)
	Effects on otters	Monson et al. (2000)
	Death due to hypothermia	Hunt (1987)

Hydrocarbon toxicity on the benthic marine community is associated with the persistent fractions of oil (NRC, 2003) as benthic species are either immotile of characterized by limited motility. The scope of growth in mussels (*Mytilus edulis*) exposed to oil for a long term showed a decline. It has been suggested that reduction in the scope of growth was related to the accumulation of two or three ring aromatic hydrocarbons due to the narcotic effect exercised on the ciliary feeding mechanisms (Donkin et al., 1990). Reduction in recruitment and over-wintering mortality in the crab *Uca pugnax* was (1995) observed as an effect of No.2 diesel oil spilled in Florida (Krebs and Burns, 1977). Disappearance of naphthalenes was followed by recovery of the crab populations. The uptake, tissue distribution and elimination of polynuclear aromatic hydrocarbons (PAH) have been studied by Meador et al. (1995). Tumors as well as other histopathological disorders were found in benthic fishes and shellfish, indicating the role of PAH in inducing histopathological conditions.

Both early and recent studies on the impact of spilled oil on seaweeds have shown that although they can suffer considerable damage, they retain their capacity to recover (Smith, 1968; Diez et al., 2009). Recent work on macroalgal communities following the *Prestige* oil spill (Diez et al., 2009) showed that macroalgal species were not seriously modified; the study could not come to a conclusion as to the possibility that spilled oil had impacted the seaweeds. There are two possible reasons explaining the results (a) the highly heterogeneous environment in species richness, diversity and cover introduces considerable variation during the statistical analysis of the data and (b) oil spill impact assessment requires prior knowledge (pre impact data) on spatial and temporal scales of species, abundance as well as on the interaction between physical, chemical and biological factors (Jackson et al., 1989; Terlizzi et al., 2005).

As fishes can move it is expected that they can keep clear of dark spills floating above them while the taste of dissolved or sinking oil can repel them from being in contact with oily masses (Nelson-Smith, 1972). This has been reported in a recent paper (Roth and Baltz, 2009); after pre and post spill sampling in the South Luisiana oil spill in Baratana Bay during 2005, it was found that fish had left the spilled area. However, oil seems to be effective on fish eggs and embryos. It has been reported high embryo mortality of pink salmon when exposed to aqueous concentrations of 1ppb of PAH during laboratory experiments (Heintz et al., 1999). Johnson et al. (2001) reported toxic effects of PAH concentrations in sediments ranging between 54 and 2800 mg/l on gonadal growth, spawning, egg viability and growth.

Marine birds can be affected by spilled oil in a number of ways (Hunt, 1987). Being air breathing organisms, seabirds have to go through the surface of the sea to catch their prey underwater. If oil is present on the surface, their feathers are coated. This can destroy the water proofing and as birds are warm blooded animals they can die from hypothermia. In addition, they can be poisoned by ingesting oil while they trying to remove it from their feathers. The upper trophic level of predators may ingest oil if their prey has been exposed to oil. Oil can also be transferred from their feathers to the eggs they incubate; oil components can affect the embryos or even the development of young birds. Location and timing of the oil spilled seem to play a crucial role to mortality rather than the size of the oil-spill (Burger, 1993). If the spill occurs when the birds are aggregated during breeding or migration, the impact is serious. Reduced reproductive success, physiological impairment and increased vulnerability to stress have been reported among the symptoms of sublethal effects of spilled oil (Fry and Addiego, 1987; Hunt, 1987; Briggs et al., 1996).

Sea mammals face similar problems as sea birds. The presence of oil on the pelage of marine mammals can affect the thermo-insulation of the animal. Marine mammals are also sensitive to ingestion of lamps of oil. There is strong evidence that spilled oil affects population dynamics of sea mammals. It has been found a population decline on African Penguins (*Spheniscus demersus*) in Southern Africa (Westphal and Rowan, 1970). However, long term effects on sea mammals are difficult to assess and a combined approach including population data, modeling techniques and statistical analysis has been proposed (Monson et al., 2000).

4. Major Oil Spill Incidence

Major oil spills in the marine environment come from oil tanker accidents and marine oil well blowouts (Table 3). The number of tanker accidents has shown a downward trend since the 1970s, followed by a reduction of the amount of oil spilled in the sea (Jernelov, 2010). Many reasons have been reported that account for this production: double hulls that limit the risk especially after minor incidence, sectioning of tanker's storage place, rules on marine traffic especially in channels and the use of GPS (Global Positioning System) for reliable navigation. The average amount of oil spilled in the sea per year from tanker accidents in 1970s was 314,000 tons whereas during the first decade of the third millennium was only 21,000 tons per year (Oil Tanker Spill Statistics, 2009). However, with regard to marine oil blowouts such a downward trend has not been observed. Blowouts in shallow waters can be handled relatively quickly and successfully but blowouts in deeper waters can go on for several months until relief wells are drilled (Jernelov, 2010) or specific techniques are innovated.

Table 3. Major oil spill incidence

Incidence	Year of incidence	Location	Oil spill characteristics	Major impacts	References
Torrey Canyon (tanker)	1967	Isles of Scilly (U.K.)	Quantity: 119,000 tons Area covered by oil: 700 km^2	Hundreds of km of beaches in Southern England and France were coated with oil. About 15,000 birds were affected by oil.	Bellamy et al. (1967) Smith (1968) Cowan (1968)
Santa Barbara (well blowout)	1969	Santa Barbara Channel in Southern California	Estimates of oil spilled during the first 10 days: 13,000-16,000 tons	A number of 3686 birds ws recorded. Deaths of marine mammals (sea lions and sea elephants) were observed. Long term effects on plankton and fish population negligible.	Ogouri and Kanter (1971)
Amoco Cadiz (tanker)	1978	Off the coast of Brittany (France)	Oil spilled: 223,000 tons	Devastating disaster on marine life. High mortality on oyster and echinoderm populations was observed. Many fishes caught showed symptoms of skin ulcers and tumors.	Osullivan (1978)
Ixtoc I Platform	1979	Bay of Campeche (Central America)	Oil spilled: 475,000 tons		Jernelov and Linden (1981) Jernelov (2010)
Exxon Valdez (tanker)	1989	Prince William Sound (Alaska)	Oil spilled: 41,000-119,000 tons Coast line affected by oil: 2,100 km Oil spill area: 28,000 km^2	Death to more than 1000 sea otters (Enhydra lutris) and about 300 harbor seals (Phoca vutulina).	Jernelov (2010)
Sea Empress (tanker)	1996	Near Pembroke (U.K.)	Oil spilled: 72,000 m^3	Affected the Pembrokeshire Coast National Park.	Moore (1996)

Table 3. (Continued)

Incidence	Year of incidence	Location	Oil spill characteristics	Major impacts	References
Erika (tanker)	1999	Brittany (France)		Marine life was killed and north-west coasts of France (Brittany) were polluted. Fishing was halted and the consumption of shellfish banned.	Laubier et al. (2004)
Prestige (tanker)	2002	Off the coast of Galicia (Atlantic)	Oil spilled: 76,000 m^3	No effects on phytoplankton were observed. Macrofaunal populations along the beaches of Spain showed significant reduction. The local fishing industry was hit.	Varela et al. (2006) Junoy et al. (2005)
Deep Horizon Plarform	2010	Near the Mississippi estuary (Gulf of Mexico)			Rooker et al. (2013)

The first large scale oil spill was from the super tanker "Torrey Canyon". Owing to a navigational error the tanker fell aground on a reef between the Cornish mainland and the Scilly Isles (U.K.). Due to lack of experience in handling large scale oil spills, it was decided to bomb the wreck and the oil by napalm. As the oil spill was cooled by the seawater underneath, only a small fraction could be burned. In addition, 10,000 tons dispersants were used to emulsify the oil and was found that detergents were far more toxic than the oil itself. The oil slick was spread to an area of 700 km^2 killing about 15,000 birds. Other short term effects on crustaceans, mollusks and fishes could not be accurately accessed but the impacts have been characterized as rather mild (Smith, 1968). Years after the Torrey Canyon spill massive amounts of green algae were found along the rocky shores. The explanation was that killing of grazers, gave the algae a period of undisturbed growth. In addition, some benthic species with planktonic larvae such as mollusks, could not settle on surfaces occupied by algae (Peterson et al., 2003).

A blowout in an underwater well while drilling happened in January 1969 in the Santa Barbara Channel near the City of Santa Barbara, Southern California. At that time it was the largest spill in U.S. and now it is classed as the third serious incident in U.S., after the Deepwater Horizon and the Exxon Valdez spills. It was estimated that within 24h since the well became wild, an area of 200 km^2 was covered by oil and within 10 days 13,000 to 16,000 m^3 of crude oil had entered the marine environment. The most serious environmental effects were on seabirds and sea mammals. The number of dead birds recorded was 3686 birds and a number of dead sea lions and sea elephants was observed. Fish populations seemed to be unaffected by the oil although a decline in some fish species was noticed. Although animal life seems to be highly sensitive to hydrocarbons, phytoplankton was rather resistant to oil. It has been reported that primary productivity in the Santa Barbara Channel, followed the normal pattern of distribution in space according to nutrient levels (Ogouri and Kanter, 1971). On a long term basis, environmental effects of the Santa Barbara oil spill seemed to be minimal; three possible explanations have been proposed: (a) the organisms in the area over tens thousands of years were used to the oil due to natural seepage in the area and they had therefore developed resistance to hydrocarbons (b) oil degrading bacteria may be thriving in such a environment rich in hydrocarbons and (c) the incidence happened between two Pacific storms that dispersed the oil over a large area.

After the Torrey Canyon oil spill, the next large scale incident by tankers was the "Amoco Cadiz" oil spill. The ship ran aground on Portsall Rocks, near the coast of Brittany (France) in March 1978. It was estimated that 219,000 tons of crude oil and about 4,000 tons of fuel oil were spilled into the sea. The "Amoco Cadiz" incident had the biggest impact on marine life ever but at the same time, it was one of the most well studied oil spills in history. Echinoderm and small crustacean populations showed a decline but they recovered after a year. It has also been reported that many species of small fishes had a strong taste of petroleum. The disaster to oyster bed was estimated to 9,000 tons. Fishes caught showed skin ulcers and tumors. About 20,000 birds died. Tidal waters and littoral ecosystems such as marshes, mudflats and sandy beaches were also heavily impacted.

A blowout at the Ixtoc I Platform in the Bay of Campeche happened in 1979. Oil was spouting from the bottom of the sea for almost nine months. The oil leaked into the water column amounted to about half a million tons. The sea depth in the area was 50m. Crabs, shrimps, squids and small fishes were severely hit. Extensive phytoplankton blooms were interpreted as the result of reduced pressure from zooplanktonic populations indicating reduction in their numbers (Jernelov and Linden, 1981). Three thousand bird casualties were reported. A species of sea turtle (L. kempii) was heavily hit by the oil as hatching places along sandy beaches had been impacted by the oil. Although five years after the incident there were still oil residues covered by sand, mussels and oysters had settled in. Fish populations recovered, may be because of the reduction on fishing pressure over a few years after the incident. Shrimps and crabs also showed rapid recovery.

The “Exxon Valdez” oil spill incident occurred in Alaska (Prince William Sound) in March 1989. It has been considered as a devastating environmental disaster due to an oil spill. Estimations on the quantity of oil spilled ranged between 41,000 and 119,000 m^3. The dispersed oil covered 2,100 km of coastline and the area of the oil slick in the ocean was about 28,000 km^2. Both short term and long term impacts are known. Short term effects included deaths of sea birds (up to 250,000 birds), sea otters (2,800 animals), harbor seals (300 seals) and orcas (22 animals). A decline in salmon populations was also observed and a great number of salmon and herring eggs were destroyed. Long term effects still exist and there are estimates that some shoreline arctic habitats can take as many as thirty years to recover. The only reaction was the response of the United States Congress to pass the Oil Pollution Act (OPA) in 1990. The OPA is discussed in the nest section of this review.

The “Sea Empress” oil spill occurred near the Milford Haven Waterway in Pembrokeshire (Wales) in February 1996. During the first week 72,000 km^3 of crude oil was dispersed into the marine environment. The spill spread near the Pembrokeshire Coast National Park. Much damage was caused to shorelines due to bulk oil, causing extensive damage to seaweed bed and invertebrates. There was also impact on cockles and other shellfish along sandy beaches. The survival rate of affected guillemots was particularly low. However, the overall effect to migratory animals was limited as the incident happened in February and most of them had not arrived in the area for breeding. Five years after the incidence the coastline had recovered and marine wildlife population levels were more or less back to normal.

The incidence with the super tanker “Prestige” happened off the coast of Galicia in November 2002. There was a serious impact on the beaches of Spain, France and Portugal as well as on the fishing industry in the area. Studies on the macrofauna, in the Galicia sandy beaches (Junoy et al. 2005), showed that beach populations were significantly reduced partly due to the clean-up operation followed after the spill. However, studies on the plankton of the N-NW Spanish coast indicated that the plankton community structure was not affected by the incident (Varela et al., 2006). In addition, the annual patterns of phytoplankton especially during the spring and summer blooms did not undergo any changes either.

The "Deepwater Horizon Platform" oil spill also known as the BP spill, occurred in April 2010 and oil was released into the sea for three months. The leak amounted to about 780,000 m^3, exceeding any oil spill that ever occurred in the U.S. Underwater plumes were reported as well as a "kill zone" around the blown well, with an estimated area of 210 km^2. The impact was not only on the ecosystem but also on fishing and tourism; both dominant Gulf's industries were seriously hit.

5. Regulatory Issues

The first assessments about the presence of oil in the sea were carried out in Britain after the 1st World War, may be due to the increasing number of ships using oil as a fuel. The British authorities eventually implemented the "Oil in Navigable Waters Act" in 1922. Later in 1926 an International Conference took place in Washington focusing at measures to decrease the amount of oil discharged in the sea. Unfortunately, in spite of the Draft Convention delivered among the parties in 1935, no agreement was achieved as some of the Member States especially Germany and Italy refused to participate to further negotiations. After the Second World War, the British Government formed a Committee in 1952, known as the "Faulkner Committee" (after the name of the person chairing the Committee) to establish a zone of no disposal around of no oil disposal, defining a zone ranging between 100 and 150 miles of the coastline and to encourage the construction of facilities to receive the oil residues from the ships. This Faulkner Committee was the preliminary stage of the Oil Pollution Convention (OILPOL). The OILPOL was the first international agreement concerning the protection of the marine environment from oil discharges from tankers. The main objectives were: (a) any oil-in-water mixture containing more than 100 ppm of oil could not be discharged in prohibited zones (b) prohibited zones were covering an area of 50 miles from the coast. The OILPOL Convention was enforced in 1958. In spite of later amendments (1971), this Convention became obsolete when the 1973/78 MARPOL Convention was enforced.

The International Convention on the Prevention of Marine Pollution by Dumping of Wastes and Other Matter, known as the "London Convention" was signed in 1972 and enforced in 1975. The Convention defines as "dumping" (Article III) "any deliberate disposal at sea of wastes or other matter from vessels, aircraft, platforms or other manmade structures at sea". According to Article IV, the dumping of wastes or other matter listed in Annex

I is prohibited. Among the harmful substances listed in Annex I, petroleum substances are included: “crude oil, fuel oil, heavy diesel oil and lubricating oils, hydraulic fluids and any mixtures containing any of these, taken on board for the purpose of dumping”.

The International Convention for the Prevention of Pollution from Ships, known as MARPOL is among the most important international conventions for the protection of the marine environment by ships from operational or accidental causes and has been updated by many amendments through the years. The Convention includes regulations aiming at preventing and mitigating marine pollution from ships and includes six technical Annexes. The regulations of Annex I refer to prevention of pollution by oil. There are limitations concerning oily discharges that are allowed under specific conditions (a) the total quantity of oil discharged into the sea should not exceed the 1/15,000 of the total cargo capacity (or 1/30,000 in the “new” oil tankers) (b) the rate of discharge should not exceed 60 liters per mile travelled by the tanker (c) no oil discharge should happen within 50 miles from the coastline and (d) “special areas” were defined characterized as sensitive and vulnerable areas. In addition, the Load-on-Top system (LOT) was recognized and further application of LOT was encouraged. The 1992 amendments to Annex I made compulsory the double hull practice for the new tankers. The last revision of Annex I was enforced in 2007.

The Convention on the Protection of the Marine Environment in the Baltic Sea that is the Helsinki Convention of HELCOM was signed by the coastal countries of the Baltic Sea. The Member States surrounding the Baltic Sea had emphasized the indispensable values of the Baltic Sea and the exceptional hydrographic and ecological characteristics as well as the sensitivity of the living resources to changes of the environment. The Convention is addressed to all sources of pollution, including oil in any form (crude oil, fuel oil, sludge, oil refuge and refined products). Annex I refers to oil of petroleum origin whereas Annexes IV and V refer to prevention of pollution from ships and offshore activities respectively.

The EEC Directive on the disposal of waste oils was a Council Directive in 1975 (Directive 75/439/EEC). Member States are obliged to take all the necessary measures to prohibit oil discharges in the marine environment (Article 4) and also to ensure that as far as possible the discharge of waste oils will be carried out by recycling (Article 3).

The Convention for the Protection of the Mediterranean Sea against Pollution (Barcelona Convention) is a regional convention aiming at preventing and mitigating marine pollution from ships, aircrafts as well as land based sources. The main objective of the Convention is to "reduce pollution in the Mediterranean Sea, protect and improve the marine environment in the area, thereby contributing to its sustainable development. Although the Member States recognize the progress achieved by the existing international conventions, they maintain that their marine pollution measures do not satisfy the special requirements of the Mediterranean Sea. Pollution from ships (Article 6) and pollution resulting from exploration and exploitation of the continental shelf and the seabed and its subsoil (Article 7) are especially mentioned in Convention. In addition, the Convention requires from the Contracting Parties to establish monitoring programs (Article 10) to monitor both pollutants and recently ecosystem quality. Scientific and technological co-operation between signatory countries are also encouraged. The competent authorities for the implementation of the scientific aspects of the Barcelona Convention are the UNEP Regional Seas Program and the Mediterranean Action Plan (MAP). The Barcelona Convention was signed in 1970 and revised in 1995 as the Convention for the Protection of the Marine Environment and the Coastal Region in the Mediterranean.

The International Convention on Oil Pollution Preparedness, Response and Co-operation (OPRC) was prepared shortly after the accidents of "Exxon Valdez" (1989) near the coasts of Alaska and "Khark-5" (1989) near the coasts of Morocco. The Convention requires from the signatory parties to establish measures for handling pollution incidents. Ships are required to carry on board an oil pollution emergency plan under the auspices of IMO. Emergency plans are also required by offshore units under the jurisdiction of the states responsible for the operation of the facilities. Ships are also required to report any incident of oil pollution to coastal authorities.

The Oil Pollution Act (OPA) was passed by the United States Congress in 1990 mainly in response to the public concern following the "Exxon Valdez" incident. The Congress consolidated the existing federal oil spill laws under one law. The law focuses at the liability of the companies to have "a plan to prevent oil spills that may occur" and have a "detailed containment and cleanup plan" for oil spills.

The Convention for the Protection of the Marine Environment of the North-East Atlantic known as OSPAR Convention (OSPAR, 2007), combines the 1972 Oslo convention on dumping waste at sea and the 1974 Paris Convention on pollution from land based sources. The Convention was signed by 15 Member States at the Ministerial Meeting of the Oslo and Paris Commissions in Paris in 1992. The Convention encourages co-operation among the signatory parties to cover "all human activities that might adversely affect the marine environment at the North-East Atlantic. The OSPAR Convention establishes standards and encourages monitoring programs to assess environmental conditions on various aspects of the environment (biodiversity, eutrophication, radioactive material) taking also care for the offshore oil and gas industry. Regulatory issues are summarized in Table 4.

Table 4. Global environmental legal instruments for the protection of the marine environment from oil pollution

Document	Year	Major environmental objectives	Specific measures on oil pollution
Oil Pollution Convention (OILPOL)	1954	It was the first International Convention attempting to prevent oil pollution by tankers in the marine environment.	Oil discharges by tankers were forbidden in prohibited areas. The convention provided stringiest measures on the load-of-top system and introduced new standards for the construction of tankers.
Geneva Convention on High Seas	1958	The convention refers to "harmful effects of oil pollution from ships and from offshore oil exploration and exploitation".	It recognizes the harmful effects of oil pollution from ships and shore activities and signatory states are obliged to take appropriate measures.
London Convention on the Prevention of Marine Pollution by Dumping of Wastes and Other Matter	Singed: 1972 Enforced: 1975	The purpose of the London Convention was the prevention of marine pollution by the dumping of wastes and other matter.	It regulates the deliberate disposal at sea of a number of substances including oily wastes. Hydrocarbons are included in Annex I (Black List) and their dumping is prohibited.
International Convention for the Prevention of Pollution from Ships (MARPOL)	Signed: 1973 Protocol: 1978 Enforced: 1983	The object of MARPOL is to prevent the pollution of the marine environment by the operational discharge of oil and other harmful substances and to minimize the accidental discharge of such substances.	An International Oil Pollution Prevention (IOPP) certificate if required from ships exceeding certain capacity. Coastal states should have installations for receiving oily wastes from ships. Annex I refers to oil pollution.
Convention on the Protection of the Marine Environment of the Baltic Sea Area	Signed: 1974 Enforced: 1980	All sources of pollution from the entire Baltic Sea Area.	"By 2013, all sub-regions of the Baltic Sea should be fully prepared to cope with medium-sized oil spills

Document	Year	Major environmental objectives	Specific measures on oil pollution
(The Helsinki Convention, HELCOM)			affecting and requiring response from more than one country. An adequate level of preparedness to respond to accidental pollution involving hazardous substances is to be achieved by 2016" (Backer et al., 2010).
Council Directive 75/439/EEC on the disposal of waste oils	1975	In this European Economic Community Directive Member States should take measures to ensure that no disposal of waste oils will be discharged in the marine environment and also encourages Member States to recycle oily waste.	Referring in detail to the handling and management of waste oils.
Convention for the Protection of the Mediterranean Sea Against Pollution (Barcelona Convention)	Singed: 1976 Enforced: 1978 Amended: 1995	The main objective is to "reduce pollution in the Mediterranean Sea and protect and improve the marine environment in the area thereby contributing to its sustainable development".	"The Contracting Parties shall take all measures in conformity with international law to prevent, abate, combat and to the fullest possible extent eliminate pollution of the Mediterranean Sea Area caused by discharges from ships.." (Article 6)
International Convention on Oil Pollution Preparedness Response and Co-operation	1990	This convention is "aiming at encouraging the development of emergency plans for oil pollution from ships and off-shore activities and at ports and coastal facilities".	Encourages international co-operation on oil pollution issues.
Oil Pollution Act (OPA) (Federal U.S. Act)	1990	Companies should "plan to prevent oil spills that may occur". Companies should have a "detailed containment and clean-up plan".	Specific requirements such as the double hulls have been introduced to mitigate the effects in case of a crash.
Convention for the Protection of the Marine Environment of the North-East Atlantic (OSPAR Convention)	Signed: 1992 Enforced: 1998	To take measures for "all human activities that might adversely affect the Marine Environment of the North-East Atlantic.	The OSPAR Convention regulates European standards on a number of human activities including the gas industry.

6. Assessment and Monitoring

Assessment of oil pollution and its impact engages a number of steps: (a) sampling strategy (b) sampling techniques from various biotas (seawater, sediment and organisms), (c) analytical procedures and (d) quantitative assessment using statistical methods and/or simulation models. Methods on sampling design for assessing oil pollution are mainly based on classical statistical methods (Cox, 1980; NRC, 1985). In addition, a number of new statistical techniques have been developed as methodological tools in water quality studies, air pollution and field sampling since 70s and they are therefore suitable for oil pollution studies (NRC, 1985). However, as oil spill incidents are not planned, it is not usually feasible to study the impact using well structured balanced designs or decide upon replication aspects (Eberhardt and Thomas, 1991). Due to the accidental nature of oil spills and the fact that do not necessarily occur in well studied areas, historical data are missing in most instances; in the case of an oil spill neither Pre/Post pair comparisons (assessment of differences between time periods based on pairs of samples from the same site) nor BACI (Before-After-Control-Impact) designs can be used. In the case of lack of historical data, only single time or multiple times experimental designs can be applied instead (Wiens and Parker, 1995). The simplest single time design is organized as an ordinary experimental treatment – control design: random samples are collected from the area where the incident occurred (samples corresponding to treatments) as well as from other reference (control) sites (Wiens and Parker, 1995). This design has already been applied to assess the effects on shoreline flora and fauna, following the *Exxon Valdez* oil spill in the summer of 1990 (Gilfillan et al., 1995). An underlying assumption in this type of design is that the effects of environmental variables are equal over both the impacted sites and the reference sites. However, this assumption should always be checked and analysis of covariance has been recommended as a useful data analysis tool (Wiens and Parker, 1995). Time series is among the most popular multiple range designs. If a set of samples collected from the impacted site at regular intervals is available, the assessor can detect temporal trends and subsequent recovery. The impact level-by-time interaction is also a type of time-course experimental design: statistically significant differences of profile means between contaminated and control sites over a period of time are assessed (Morrison, 1988). A third multiple time design method is the impact trend-by-time interaction: if variables in the oil polluted area are measured as continuous variables, possible trends between measures of the resource and exposure level over time can be revealed. Single time designs, seem to suit rather to surveys for the impact assessment, whereas multiple time designs provide information to temporal dynamics and therefore seem to suit more to trend monitoring programs.

Analytical procedures can take place in the field using *in situ* fluorometers provided with a continuous pumping device on board. In addition, information collected using simple devices to determine the water content of oil, to assess the emulsification state and to measure the specific gravity, are useful to deplore operational plans including use of dispersants, application of booms as well as to estimate the clean up efficiency (NRC, 1985).

Behavior, environmental fate and biological/ ecosystem effects of spilled oil depend on the physical and chemical properties of the material. It is therefore necessary to delineate oil composition. A number of techniques are available at laboratory level: (a) gas chromatography (GC) provides considerable information on the molecular weight of the hydrocarbon components of the sample; GC has been in use since 60s and it is still considered as a powerful tool in the characterization of crude oils for both the paraffin fraction (Crowley et al., 1980; Clark and Jurs, 1979) and the aromatic components (Atlas , 1981). Use of Infrared Spectrometry (IR) is applied to characterize major compound groups. IR can also provide information on weathering processes by detecting carboxyl and hydroxyl functional groups (Blumer et al., 1973). Mass Spectrometry (MS) techniques provide quantitative information on almost twenty five molecular types that have been found very useful in evaluating the chemical characteristics of fresh and weathered oils (Robinson and Cook, 1969; ASTM, 1980). Detailed composition analysis of oils has been attained using GC/MS techniques (Albaiges and Albrecht, 1979; Hood and Erickson, 1980). HPLC is used for characterizing oils according to their aromatic hydrocarbon content. Combination of IR and HPLC methods has provided quantitative information on saturated and aromatic petroleum hydrocarbons. The Ultraviolet-Visible Fluorescence (UV/F) Spectrometry also known as spectrofluorometry is widely used for the determination of background levels of petroleum hydrocarbons from seawater, sediments and biological material (Keizer and Gordon, 1973; Zitko, 1975; Fong, 1976; Ignatiades and Mimicos, 1977; Wakeman, 1977; Karydis, 1982b).

Hydrocarbon analysis from a particular sample of water, sediment or tissue assumes prior recovery of petroleum compounds from the sample. There is a number of techniques for each kind of biota. The procedure to be selected depends on: (a) the kind of biota (b) the hydrocarbon components the assessor is interested in and (c) the analytical methodology to be used. Description of recovery methods for studying oil effects at physiological, behavioral, population and ecosystem levels are given in NRC (1985).

There are many adverse environmental effects should an oil spill incident happen. In addition to the initial survey, it is equally important to monitor the incident. On a short term basis monitoring of the oil slick itself provides information about the slick spreading and the fate of oil in the aquatic environment. Long term monitoring is necessary to assess the effects on organisms and the ecosystem. This type of monitoring is known as trend monitoring (UNEP, 1997). A monitoring program of trends records the levels of hydrocarbons over the years and can provide quantitative information on the downward trend following the incident as well as information concerning changes on the chemical composition of petroleum compounds due to weathering effects. At the same time sampling on the biota can show possible biological effects over the time and ecosystem recovery. A well planned monitoring program should take into account the following (UNEP, 1997): (a) Clear objectives. Oil pollution monitoring programs over the last three decades address two main targets (NRC, 1985): detection of environmental changes due to petroleum hydrocarbons and assessment of recovery of an oil spill stressed system. In particular a trend monitoring program should set confidence intervals for each variable measured so that the minimum trend can be identified (b) optimal station selection. The station selection should satisfy the managerial objectives of the program. The sites should include a sufficient number of biota required for the monitoring program and also the site should be suitable for sediment clam-core analysis (c) well defined sampling and analytical procedures so that sampling requirement, effort and cost can be calculated. Assessment of polynuclear aromatic hydrocarbons and in water and biota are recommended for oil pollution monitoring (UNEP, 1997). When the amounts of oil are large, remote sensing seem to be the optimal approach. For low levels of hydrocarbons, bench work is required. The UV/F method can be applied and "equivalent oil concentration" can be obtained (NRC, 1985). If individual compounds should be indentified or even a single class of compounds (e.g. aromatic hydrocarbons or n-paraffins), the determination is more expensive and extensive quality control/ intercalibration exercises are required (UNEP, 1997; NRC, 1985), (d) selection of organisms to be monitored; species selection should take into account routine laboratory practices as well as the ecological and/or commercial value of the organisms. However, focusing on community structure (community assemblages), species composition, species succession and biomass determinations are recommended

(NRC, 1985). An Environmental additional characteristic of the ecosystem that should be examined together with the recovery is the "environmental capacity". Environmental capacity is defined "as the property of the environment and can be considered as the ability to accommodate a particular activity or rate of activity without unacceptable impact" (UNEP, 1997), (e) Timing and frequency of sampling.

7. REMOTE SENSING

In recent years, airborne and satellite remote sensing observations are considered as the most important tools for detecting oil spills at sea surface. Identification of oil spills in the marine environment can be carried out by multiple sensors by comparing fundamental characteristics of oil to those of water, such as reflection, emissivity or conductivity (Estes et al., 1972). The basic principle is the detection of electromagnetic radiation and enhancement of the contrast between the 'object' and the background (Martin 2004). Naturally reflected and/or emitted electromagnetic radiation is recorded by passive sensors, while active approaches include Light Detection and RAnging (LiDAR) and radar.

High resolution microwave observations provided by Synthetic Aperture Radars (SAR) and datasets acquired from sensors operating in the visible/near-infrared (VNIR) and thermal regions of the electromagnetic spectrum are used for oil spills' identification. Since VNIR techniques are based on the detection of the upwelling radiance to discriminate between clean and oil polluted waters, the impact of the atmosphere is significant. Therefore, accurate atmospheric correction is needed in order to remove the contribution of the atmosphere to the radiance registered to sensor. Remote sensing in Thermal InfraRed (TIR, 7000–14,000 nm) focuses on the identification of thermal contrast spatial patterns in order to discriminate oil from seawater. The thermal contrast arises from the different emissivities of water and oil depending on slick thickness (Salisbury et al., 1993) as well as the oil's lower heat capacity and different thermal conductivities.

MODIS (Terra and Aqua satellites) and MERIS (European Space Agency) multispectral instruments can identify oil spills in the visible spectrum based on sunglint and the fact that slicked surfaces have higher reflectance than pure water. Adamo et al. (2009) compared MERIS to MODIS datasets in the Mediterranean Sea and found better slick discrimination for longer wavelengths with greater sunglint sensitivity for MERIS. The MODIS provides rapid revisit time (twice daily) compared to MERIS (3 days). De Carolis et al. (2014) investigated the capability to retrieve the average thickness of marine oil slicks by using the NIR bands of the MERIS and MODIS sensors in the Lebanon oil spill case study. Bulgarelli and Djavidnia (2012) used also MODIS datasets to retrieve oil spill spectral properties in the marine environment. They concluded that the detection of oil features in the absence of sunglint contamination is difficult using MODIS data; there is low visibility of oil spills outside the glint zone, in particular for small spills.

The Landsat 5 TM and Landsat 7 ETM+ sensors have been used to map oil slicks as well. They provide significantly higher spatial resolutions, but a relatively infrequent revisit (about 16 days), making them unsuitable for slick monitoring (Bentz and Pellon de Miranda, 2001).

Airborne and satellite SAR datasets are widely used for oil slick detection (Brekke and Solberg, 2005; Fingas & Brown, 2011). Radar spectral section is divided in several bands; for oil spill tracking, Ku, X, C, S and L bands (1.7 cm $\leq \lambda \leq$ 30 cm) are used. The principle is the identification of changes in surface wave properties as a result of oil dampening, since oil dampens the intensity of the high-frequency, wind-generated sea waves. Oil reduces the surface tension and the wind friction effect on the sea surface; thus, reduction and shift of short gravity and capillary waves to shorter wavelengths appear along with an effective reduction of wave growth (Leifer et al., 2012). As a result, both the sea radar backscatter and the energy received by the sensor are reduced. Therefore, an oil spill appears as a dark area in an intensively bright background (Brekke and Solberg, 2005; Di Giacomo et al., 2004).

Radar images alone do not provide sufficient information for the identification of oil spills, since several types of 'look-alikes' can appear (Topouzelis et al., 2007). For different radar bands, the threshold wind speeds for significant Bragg backscatter depend on radar frequency, incidence angle and sea surface temperature; very low wind speeds (< 2 m/s) are not able to roughen the sea surface through capillary wave formation, therefore the sea surface is flat (Donelan and Pierson, 1987). Calm seas represent reduced radar backscatter similar to backscatter of an oil-slicked surface, making slick detection impossible (Trivero and Biamino, 2010). The ideal range of wind speed is between 2-3 m/s and 12 m/s, with the optimum around 5-6 m/s (Torres Palenzuela et al., 2006).

SAR sensors have been widely used in oil spills' studies since they can be used throughout the day (24 h/day), without being affected by weather or clouds. At the same time, computing power has been dramatically increased allowing the development of highly sophisticated algorithms for image processing (Topouzelis et al., 2006). They have also been quite successfully used for oil seepage detection, during the exploration for oil sources (O' Brien et al., 2005). The effectiveness of a satellite SAR system depends on the strength of the produced contrast signal and on the area size in which the system can reach adequate coverage. Furthermore, both depend on oil type and age (Pavlakis et al., 1996), that is the temporal space between spillage and detection (Brekke and Solberg, 2005). All physicochemical processes that modify the basic oil spill properties, as well as the pace in which they occur, depend on oil spill age. Many studies have proved the effectiveness of SAR for oil spill monitoring (Shi et al., 2008; Vespe et al., 2011; Kontoes et al., 2005). SAR images acquired by the European Remote Sensing (ERS)-1/2, Envisat and Radarsat satellites have been extensively used to assess oil pollution in various marine areas (Pavlakis et al., 1996) frequently combined with geo-information approaches Miranda et al. (2004).

8. Geographical Information Systems (GIS) and Modeling

Many of the numerical models developed to study various aspects of oil pollution (dispersion, fate, processes, ecological impact) can also be used for monitoring, although they have not been designed for long term monitoring. Some of the models have been designed to develop maps indicating the probability of oil spill dispersion and its impact on the region of the spill (NRC, 2003). Models can also be used for post-spill assessment. The idea is to fill observation gaps and estimate damage. Simulation models used for the monitoring of an oil slick are complex due to oil weathering: evaporation, emulsification, dissolution, photo-oxidation and biological biodegradation change oil composition, bioavailability and toxicity with time and those processes should be included in the model design. Luckily, the interdependence of weathering processes in oil spills have been known for a long time (NRC, 1985). There are two rather extensive reviews on composite oil spill models by ASCE (1996) and Reed et al. (1999). The OSRA model (Smith et al., 1982) is applied to assess spill impact probability related to offshore oil development in the Outer Continental Shelf. The model SIMAP (French, 1998) is mainly used for post-spill damage assessment; the model calculates both fates and effects. Reed et al. (1999) reviewed numerical models that referred on behavior, fate and biological effects of oil on marine organisms and biota. Effects of oil pollution are closely related to the natural dispersion of an oil spill. Monitoring oil dispersion is a necessary step in order to access (a) the area of dispersion (b) the lifetime of an oil spill and (c) biological effects. The rate of dispersion and fate depend not only on environmental parameters (wind speed, temperature and light intensity) but also on oil characteristics (density, surface tension and viscosity). Among many models on dispersion the work of Delvigne and Sweeney (1988) is the most common in use. The basics of this methodology are now incorporated in more complex models: the model ADIOS (NOOA, 1994) simulating detailed weathering and dispersion; the SINTEF model (Daling et al., 1997) as well as the OILMAP are models describing weathering. The fate of oil spill at sea is also simulated by a model described by Sebastiao and Soares (1995). Slick dispersion has also been modeled by other authors (Audunson et al., 1980; Blikely et al., 1977; Liu and Leendertse, 1986). An important issue on marine monitoring studies in respect to oil pollution is the oil impact on the shoreline. Nowadays, there are several models that include dynamic processes of oil in the coastal zone. The coastal zone oil spill model COZOIL (Reed et al., 1989) simulated oil sediment interactions, wave propagation and a representation of the shoreline which is described segment by segment. This model has been tested with data from the Amoco Cadiz oil spill (Reed et al., 1989). However, oil pollution problems especially of chronicle type, occur in hot spot coastal areas near harbors and oil terminals (EEA, 1999). This problem raises the need for model development that would link oil pollution to other pollutants. This type of modeling will be particularly useful in coastal areas facing a number of environmental pressures.

GIS is used as a framework for (i) the application of models and the visualization and mapping of their output illustrating predictive spill fate and transport, (ii) the development of spatial databases and the application of spatial analysis and mapping (iii) the evaluation of oil spill environmental sensitivity which depends on the degree of impact the oil can exert on an area and on its persistence and (iv) the development of decision-support systems. During the last years, there is a rapid development of webGIS applications, a hybrid of GIS and Internet technologies, allowing the interaction of users with spatial databases over the web. Kulawiak et al. (2010) developed a web-based GIS system for the interactive visualization of oil-spill monitoring and forecasting. The latter was achieved by using the operational tool for prediction of floating pollutant transport of the POSEIDON system (Pollani et al., 2001), which consists of two discrete modules: (i) a three-dimensional oil-spill weathering and drift module that simulates the dispersion of oil droplets and their chemical transformations, and (ii) a forcing module that provides data for wind, waves, currents and diffusivity to the oil–drift module. The advantage of this approach is the access of the end-users via a user-friendly graphical interface to the animated visualization of the oil spill spread in combination with tools for querying each hour layer for parameters like the volume of selected oil spill, its geographic coordinates etc.

SMIS (Spill Management Information System) is a GIS-based decision support system with live real-time data links designed to manage the risks associated with releases of hazardous material (Martin et al., 2004). A Database Management System (DBMS) is coupled with the 2D surface water model CE-QUAL-W2 and the air contaminant model (CAMEO) in a GIS framework. The establishment of SMIS as part of a decision support organization could assist environmental response teams to a chemical spill emergency. An improved version of SMIS to inland waterways has been proposed by Camp et al. (2010); it provides improved predictive spill fate and transport capability combined with a GIS environment to assist decision support for planning and response activities. Within this framework, the 3D Generalized, Longitudinal-Lateral-Vertical Hydrodynamic and Transport (GLLVHT) model is applied for providing hydrodynamic information for contaminant transport through the Chemical/Oil Spill Impact Model (COSIM). Teruhisa et al. (2003) developed a GIS approach for monitoring and analyzing changes in an intertidal ecosystem polluted by heavy oil at a spatial scale of 10–100 m along the shore and at an intermediate temporal scale. GIS proved able to trace temporal changes in intertidal communities resulting from the impacts of heavy oil on flora and fauna. Concerning oil spills sensitivity mapping, Krishnan (1995) demonstrated the efficiency and effectiveness of the GIS approach in determining the critical areas which should be protected in case of an oil spill. Olita et al. (2012) proposed a model based method to evaluate hazard of oil slicks contact with shorelines of Sardinia. The approach involved application of hydrodynamic models and the development of a hazard index in the framework of a GIS applied to assess and map the risk from oil dispersal at sea. An expert GIS system to evaluate oil spill environmental sensitivity has been proposed by Pincinato et al. (2009). The system works based on rules set by the official standards of Brazilian Federal Environment Organ and employs a decision tree to assign oil sensitivity indices to Brazilian coasts. The usefulness of the approach is the prioritization of areas for direct intervention compared to others, since the prediction of the potential behavior of spilled oil towards the habitats is enabled. Another effort to map a socioeconomic and natural vulnerability index for oil spills has been described by Nogueira de Andrade et al. (2010), since vulnerability studies are considered as the essential first step for integrated coastal management. The role of GIS in this approach proved very important in order to carry out spatial analysis in the study area. In a similar study, Castanedo et al. (2009) integrated physical, biological and socio-economic aspects to map coastal oil spill vulnerability; this integration is considered important since all the above mentioned aspects are basic in coastal management approaches. Moreover, risk assessment maps of oil spill for major desalination plants have been developed by Elshorbagy and Elhakeem (2008) based on numerical simulations for hypothetical oil spills based on extreme case conditions. For this purpose, coupling of the hydrodynamic model (Mike3-HD) and the oil spill model (Mike3-SA) has been carried out. The source to destination travel time of an oil slick at any time is directly spotted and the associated critical wind direction is pointed out by locating the spill source on the contour maps; these maps are of practical importance for the operators of the desalination plants and the water authorities. Combination of GIS and multivariate statistical methods in the post-processing of the application of the Oil Spill Risk Analysis Model (OSRAM) output proved a powerful tool for the identification of areas of risk from multiple spill sources (Guillen et al., 2004). OSRAM is used to predict the movement of potential oil spills greater than 1000 bbl which originate from offshore oil and gas facilities.

9. Discussion

Oil spills can be one of the most damaging threats for coastal environment. Their hazardous environmental and socio-economic impact makes the need for the development of appropriate prevention and management tools crucial.

Monitoring of oil pollution has been improved due to applications of remote sensing. Satellites can play a significant role in oil spill detection; however, severe limitations arise from overpass frequency and timing and requirements for clear daytime sky for passive sensors. In addition, of significant concern is the spatial resolution, i.e. MODIS with a spatial resolution of approximately 1 km. Since oil spills are in general of small-scale, they cannot be detected from coarse resolution sensors. On the other hand, sensors with finer spatial resolutions but infrequent narrow swath data could miss slick detection as well. Nested data collection could be a solution as the finer scale sensor could provide sub-pixel information for the synoptic view sensor (Leifer et al., 2012). Temporal resolution of sensors is also a limitation, since oil slicks are highly dynamic and evolve on daily to hourly time-scales losing most useful information in less than a day. Therefore, there is no meaning of mapping oil spills in subsequent multi-day overpasses of the satellite.

Future passive optical sensors would become more effective to detect oil spills if higher spectral and spatial resolution combined with higher SNR is achieved. Therefore, they could have a significant contribution to operational services (Bulgarelli and Djavidnia, 2012). Comparing airborne to satellite remote sensing it seems that airborne sensors have a number of advantages, since (a) each oil spill is unique; therefore, easy application of remote sensing without significant human interpretation is prevented and (b) direct visual airborne observations can provide observations of the oil spill over a wider range of conditions (Leifer et al., 2012).

Improvement of the remote sensing techniques to discriminate between oil-slicks and look-alikes based on pattern recognition will increase the reliability of the method. Neural network approaches can discriminate slick-like features from other dark, non-slick features (Topouzelis et al., 2007). In addition, spill identification can be supported by consideration of ancillary data. This improvement will help oil-spill modeling in several ways (Reed et al. 1989). Coupling an oil-spill model to an ecological impact model provides an objective evaluation of the Net Environmental Benefit of Alternate (NEBA) to consider different oil-spill response strategies (Reed et al., 1999) when contingency planning is designed. There are also oil pollution models simulating changes on long term basis, referring to biological impact (Fucik and Show, 1981). The contribution of drifter buoys (or drifters) on oil spill detection is also significant, since they are used for the indirect calculation of various parameters in the marine environment, including oil spills (Goodman, 1994).

GIS can contribute significantly to the assessment and monitoring of oil pollution in combination with remote sensing and modeling techniques. GIS can provide the framework for the application of integrated approaches which have been proved important for operational services and include: (a) The development of accurate and easily updated spatial databases where spatial and attributes information from the study area (meteorological, hydrographic, climatic, satellite, historical, real-time data and field observations) can be stored and processed. (b) The elaboration of comparative geospatial analysis of the data in order to create multilayer maps. (c) The combination with remote sensing techniques in order to locate objects and characterize regions of interest in satellite scenes, improving therefore oil spill identification and classification of SAR imagery (Ivanov and Zatyagalova, 2008). (d) The application of spatial analysis methods for the development and mapping of environmental sensitivity indices (Jensen et al., 1990) and oil spill hazard indices (Garcia et al., 2013) for assessing and mapping the vulnerability (environmental and socio-economic) of the threatened areas; Romero et al. (2013) proposed an integrated assessment for developing an oil environmental vulnerability map by combining numerical modeling, information about environmental sensitivity to oil, application of scenarios and mapping techniques. (e) The integration and application of modeling components and mapping of their output results. (d) The providing of a user friendly interface, usually a webGIS interface for application of the integrated methodologies, the development and application of scenarios in order to support decision making and risk management. The development of operational systems and services for monitoring and forecasting coastal and marine oil pollution by providing near real-time information assessed in a user-friendly way by the relevant authorities is essential and a prerequisite for effective monitoring and decision-making. The operational systems for response to oil spill support the agencies related to marine safety by adopting measures to prevent or reduce the catastrophic impact on the marine environment. The operational response to oil spill incidents includes the identification of the exact location of the spill, the prediction of its disperse direction, the potential location and time of its arrival to the coast, the type of resources and ecosystems that will be threatened and the oil spill properties after a given time.

The importance of integrated oil spill management systems has been revealed by many researchers (Wirts and Liu, 2006; Assilzadeh and Gao, 2010). The development and application of a hybrid ocean forecasting system for oil spill in the Mediterranean coastal zone (Jordi et al., 2006) consists of three inter-connected modules: (i) an operational circulation sub-system, (ii) an oil spill coastal sub-system and (iii) a risk management sub-system for supporting decision-making based on GIS. Different forcing scenarios have been applied and the results indicated the importance of such integrating approaches to oil spill response at different space scales. The well-established MEDSLIK oil spill and trajectory prediction system (Zodiatis et al., 2012) is adapted to the MyOcean regional products and supports the relevant agencies with important information related to marine safety. In addition, MEDSLICK offers several tools to estimate where the oil spill originated, in order to trace its path backwards in time (hindcast). Such a tool may be of use following the

EU Directive 2005/35 in order to assign the costs of cleaning it up and any damage it may have caused.

The development of integrated dynamic user-friendly systems such as webGIS (Fustes et al., 2014), incorporating economic, social, environmental and ecological priorities and risk analysis can further support direct communication and interaction among authorities, stakeholders and the wider public, which is essential in the case of an emergency situation. The parties involved are able to rationalize their views and adopt effective policy making decisions relying on sound science. In the webGIS systems mapping can be updated periodically to comply with the current situation and local changes in the study area, facilitating therefore risk analysis performance and investigation of the potential impact of alternative policy scenarios on the threatened areas. This information is disseminated and shared among scientists, stakeholders and the public facilitating their collaboration and exchange of knowledge. Risk analysis, incorporating risk assessment and risk management, is also a prerequisite for spatial planning (Santos et al., 2013); since oil spills can cause significant impacts in marine and coastal environments the development of preparedness and response tools is important for sustainable management in the framework of international policies for the reduction of marine pollution (OSPAR and MARPOL Conventions).

REFERENCES

Adamo, M., De Carolis, G., De Pasquale, V. and Pasquariello, G. (2009). Detection and tracking of oil slicks on sun-glittered visible and near infrared satellite imagery. *International Journal of Remote Sensing,* 30, 6403–6427.

Albaiges, J. and Albrecht, P. (1979). Fingerprinting marine pollutant hydrocarbons by computerized gas chromatography-mass spectrometry. *International Journal of Environmental Analytical Chemistry,* 6, 171–190.

Anderson, A.G. (2002). The media politics of oil spills. Spill Science and Technology Bulletin. 7, 7-15.

API (1963). *Manual on disposal of refinery wastes: waste water containing oil.* American Institute of Petroleum, 7th ed., New York.

ASCE (1996). State-of-the-art Review of modeling transport and fate of oil spills, ASCE Committee on Modeling Oil Spills, Water Resources Engineering Division. *Journal of Hydraulic Engineering,* November, pp. 594-609.

Assilzadeh, H. and Gao, Y. (2010). Designation of an interactive oil spill management system. *Disaster Prevention and Management,* 9(2), 233-242.

ASTM, American Society for Testing and Materials. (1980). ASTM Method D 3239-76. Aromatic types analysis of gas-oil aromatic fractions by high ionizing voltage mass spectrometry. *Annual Book of ASTM Methods,* Book 25.

Atlas, R.M. (1981). Microbial degradation of petroleum hydrocarbons: an environmental perspective. *Microbiological Reviews*, 45, 180–209.

Atlas, R.M. and Bartha, R. (1987). *Microbial Ecology.* Benjamin/Cummings Publishing, Menlo Park, CA.

Atlas, R.M. and Bartha, R. (1992). Hydrocarbon biodegradation and oil spill bioremediation. K.C. Marshall (Ed.). *Advances in microbial Ecology.* Plenum Press, New York, pp.287-238.

Audunson, T., Dalen, V., Mathison, J.P., Haldorsen, J. and Krogh, F. (1980). SLICKFORCAST: a simulation program for oil spill emergency tracking and long-term contingency planning. *Proceedings Petromar 1980,* Monaco, May 26-30. Continental Shelf institute.

Backer, H., Leppanen, J.-M., Brusendorff, A.C., Forsius, K., Stankiewicz, M., Mehtonen, J., Pyhala, M., Laamanen, M., Paulomaki, H., Vlasov, N. and Harranen. T. (2010). HELCOM baltic Sea Action Plan - A regional programme of measures for the marine environment based on the Ecosystem Approach. *Marine Pollution Bulletin,* 60, 642-649.

Batten, S.D., Allen, R.J.S. and Wotton, C.O.M. (1998). The effects of the Sea Empress oil spill on the plankton of the Southern Irish Sea. *Marine Pollution Bulletin,* 36(10), 764-774.

Bellamy, J., Clarke, P.H., John, D.M., Jones, D. and Whittick, A. (1967). Effects of pollution from the Torrey Canyon on Littoral and Sublittoral Ecosystems, *Nature,* 216, 1170-1173.

Bentz, C. and Pellon de Miranda, F. (2001). Application of remote sensing data for oil spill monitoring in the Guanabara Bay, Rio de Janeiro, Brazil. *International Geoscience and Remote Sensing Symposium,* 2001. IGARSS '01 (pp. 333–335). Sydney, Australia: IEEE.

Berridge, S.A., Thew, M.T. and Loriston-Clarke, A.G. (1968). The formation and stability of emulsions of water in crude petroleum and similar stock. Pp. 35-59 in: *Scientific aspects of pollution of the sea by oil.* Institute of Petroleum, London.

Blumer, M., Ehrhardt, M. and Jones, J.H. (1973). The environmental fate of stranded crude oil. *Deep Sea Research,* 20, 239-260.

Brekke, C. and Solberg, A.H.S. (2005). Oil spill detection by satellite Remote Sensing (review). *Remote Sensing of Environment,* 95, 1-13.

Briggs, K.T., Yoshida, S.H. and Gershwin, M.E. (1996). The influence of petrochemicals and stress on the immune system of seabirds. *Regulatory toxicology and Pharmacology*, 23, 145-155.

Brookman, G.T., Flanagan, M. and Kebe, J.O. (1985). *Literature survey: hydrocarbon solubilities and attenuation mechanisms.* American Petroleum Institute, Publication Number 4414, Washington, D.C.

Buhler, D.R. and Williams, D.E. (1988). The role of biotransformation in the toxicity of chemicals. *Aquatic Toxicology,* 11, 19-28.

Bulgarelli, B. and Djavidnia, S. (2012). On MODIS Retrieval of Oil Spill Spectral Properties in the Marine Environment. *IEEE Geoscience and Remote Sensing Letters,* 9(3), 398-402.

Burger, A.E. (1993). Estimating the mortality of seabirds following oilspills; effects of spill volume. *Marine Pollution Bulletin,* 26, 140-143.

Burgherr, P. (2007). In-depth analysis of accidential oil spills from tankers in the context of global spill trends from all sources. *Journal of Hazardous Materials,* 140, 245-256. 265pp.

Butler, G.C. (Ed.). (1978). Principles of ecotoxicology. Scientific Committee on Problems of the Environment. *J. Wiley and Sons,* NY. 350pp.

Camp, J.S., LeBoeuf , E.J. and Abkowitz, M.D. (2010). Application of an enhanced spill management information system to inland waterways. *Journal of Hazardous Materials,* 175, 583–592.

Castanedo, S., Juanes, J.A., Medina, R., Puente, A., Fernandez, F., Olabarrieta, M. and Pombo, C. (2009). Oil spill vulnerability assessment integrating physical, biological and socio-economical aspects: Application to the Cantabrian coast (Bay of Biscay, Spain). *Journal of Environmental Management,* 91, 149–159.

Clark, H.A. and Jurs, P.C. (1979). Classification of crude oil gas chromatograms by pattern revognition techniques. *Analytical Chemistry,* 51(6), 616-623.

Cowan, E. (1968). *Oil and water – The Torrey Canyon Disaster.* Lippincott, Philadelphia.

Cox, C.P. (1980). *A handbook of introductory statistical methods.* Wiley, New York.

Crowley, R.J., Siggia, S., Uden, P.C. (1980). Class Separation and Characterization of Shale Oil by Liquid Chromatography and Capillary Column Gas Chromatography. *Analytical Chemistry,* 52(8), 1224-8.

Dahl, E., Laake, M., Tjesse M.K., Eberlein, K. and Boehle, B. (1983). Effects of Ecofisk crude oil on an enclosed planktonic ecosystem. *Marine Ecology Progress Series,* 14, 81-91.

Daling, P.S., Aamo, O.M., Lewis, A. and Strom-Kristiansen, T. (1997). SINTEF/IKU oil weathering model: predicting oil properties at sea. In: *Proceedings1997 Oil Spill Conference.* API Publication No. 4651, Washington DC, pp.297-307.

Davis, S.J. and Gibbs, C.F. (1975). The effect of weathering on a crude oil residue exposed at sea. *Water Research,* 9, 275-285.

De Carolis, G., Adamo, M. and Pasquariello, G. (2014). On the Estimation of Thickness of Marine Oil Slicks From Sun-Glittered, Near-Infrared MERIS and MODIS Imagery: The Lebanon Oil Spill Case Study. *IEEE Transactions on Geoscience and Remote Sensing,* 52(1), 559-573.

Delvigne, G.A.L. and Sweeney, C.E. (1988). Natural dispersion of oil. *Oil and Chemical Pollution,* 4, 281-310.

Di Giacomo, P.M., Washburn, L., Holt, B. and Jones, B.H. (2004). Coastal pollution hazards in Southern California observed by SAR imagery: stormwater plumes, wastewater plumes and natural hydrocarbon seeps. *Marine Pollution Bulletin,* 49(11/12), 1013-1024.

Diez, I., Socilla, A., Santolana, A. and Gorostiaga, J.M. (2009). Ecological monitoring of intertidal macrobenthic communities on the Basque Coast (N. Spain) following the Prestige oil spill. *Environmental Monitoring and Assessment,* 159, 555-575.

Dodd, E.N. (1971). *Report of working party on the effects of natural factors on the movements, dispersal and distribution of oil at sea.* Ministry of Defence (Navy Department).

Donelan, M. A. and Pierson, W. J. (1987). Radar scattering and equilibrium ranges in wind-generated waves with application to scatterometry. *Journal of Geophysical Research,* 92(C5), 4971–5029.

Donkin, P., Widdows, J., Evans, S.V., Worrall, C.M. and Carr, M. (1990). Quantitative structure –activities relationships for the effects of hydrophobic chemicals on rate of feeding by mussels (Mytilus edulis). *Aquatic Toxicology,* 14, 277-294.

Eberhardt, L.L. and Thomas, J.M. (1991). Designing environmental field studies. *Ecological Monographs,* 61(1), 53-73.

EEA (1999). State and pressures of the marine and coastal Mediterranean environment. Copenhagen: European Environment Agency.

Elshorbagy, W. and Elhakeem, A. (2008). Risk assessment maps of oil spill for major desalination plants in the United Arab Emirates. *Desalination,* 228(1-3), 200-216.

Estes, J.E., Senger, L.W. and Fortune, P.R. (1972). Potential applications of Remote Sensing techniques for the study of marine oil pollution. *Geoforum,* 9, 69-81.

Fay, J.A. (1969). *The spread of oil slicks on a calm sea;* pp. 53-63 in Oil on the sea. Plenum Press, New York.

Fingas, M.F. (1995). A literature review of the physics and predictive modeling of oil spill evaporation. *Journal of Hazardus Materials,* 42, 157-175.

Fingas, M.F., Fieldhouse, B., Lane, J. and Mullin, J.V. (2000). Studies of water-in-emulsions: long term stability, oil properties and emulsions formed at sea. *Proceedings of the Twenty Third Arctic and Marine Oil Spill Program Technical Seminar. Environment Canada,* Ottawa, Ontario, pp. 145-160.

Fingas, M., and Brown, C. E. (2011). Oil spill remote sensing: A review. *Oil spill science and technology* (pp. 111–169). Boston: Gulf Professional Publishing.

Fong, W. C. (1976). Uptake and retention of Kuwait Crude oil and its effects on oxygen uptake by the soft-shell clam, *Mya arenaria. Journal of the Fisheries Research Board of Canada*, 33, 2774-2780.

French, D.P. (1998). *Final Draft Report: Estimate of Injuries to Marine Communities Resulting from the North Cape Oil Spill Based on Modeling of Fates and Effects,* Report to NOAA Damage Assessment Center, Silver Spring, MD, USA, January 26.

Fry, D.M. and Addiego, L.A. (1987). Hemolytic anemia complicates the cleaning of oil seabirds. *Wildlife Journal,* 10(3), 3-8.

Fucik, K.W. and Show, I.T. (1981). *Environmental synthesis using an ecosystems model.* Pages 329-353, In B.S. Middleditch (ed.), Environmental Effects of Offshore Oil Production. The Buccaneer Gas and Oil Field Study. Plenum Press, New York.

Fustes, D., Cantorna, D., Dafonte, C., Arcay, B., Iglesias, A. and Manteiga, M. (2014). A cloud-integrated web platform for marine monitoring using GIS and remote sensing. *Application to oil spill detection through SAR images. Future Generation Computer Systems,* 34, 155–160.

Garcia, D.A., Bruschi, D., Cumo, F., and Gugliermetti, F. (2013). The Oil Spill Hazard Index (OSHI) elaboration. An oil spill hazard assessment concerning Italian hydrocarbons maritime traffic. *Ocean and Coastal Management* 80, 1-11.

Gilfillan, E.S., Page, D.S., Boehm, P.D. and Harner, E.J. (1995). Shore-line ecology program for Prince William Sound, Alaska, following the Exxon Valdez oil spill. Part 3: biology. In: Wells PG, Butler JN, Hughes JS (eds) Exxon Valdez oil spill: fate and effects in Alaskan waters. *American Society for Testing and Materials Special Technical Publication* No. 1219. ASTM, Philadelphia, PA, pp 398–443.

Girling, A.E. (1989). Preparation of aqueous media for aquatic toxicity testing of oils and oil based products: a review of the published literature. *Chemosphere,* 19(10/11), 1635-1641.

Gonzalez-Meler, M.A., Blanc-Betes, E., Flower, C.E., Ward, J.K. and Gomez-Casanovas, N. (2009). Plastic and adaptive responses of plant respiration to changes in atmospheric CO2 concentration. *Physiologia Plantarum,* 137, 473-484.

Goodman, R. (1994). Overview and future trends in oil spill Remote Sensing. *Spill Science and Technology Bulletin,* 1(1), 11-21.

Guillen, G., Rainey, G. and Morin, M., 2004. A simple rapid approach using coupled multivariate statistical methods, GIS and trajectory models to delineate areas of common oil spill risk. *Journal of Marine Systems,* 45 (3,4), 221-235.

Heintz, R.A., Short, J.W. and Rice, S.D. (1999). Sensitivity of fish embryosto weathered crude oil: Part II. Increasaed mortality of pink salmon (*Onchochyncus gorbuscha*) embryos incubating downstream from weathered Exxon Valdez crude oil. *Environmental Toxicology and Chemistry,* 18, 205-216.

Hibbs, D.E., Gulliver, J.S., Voller, V.R. and Chen, Y.H. (1999). An aqueous concentration model for riverine spills. *Journal of Hazardous Materials,* 64A, 37-53.

Holland, G. and Pugh, D. (Eds). (2010). *Troubled waters. Ocean Science and Governance.* Cambridge University Press. 316pp.

Hood L.V.S. and C.M. Erikson (1980). *Analysis of Alaskan crude oils by glass capillary gas chromatography*/mass spectrometry. Journal of High Resolution Chromatography, 3(10), 516–520.

Hook, L.A., Franco, P.J. and Giddings, J.M. (1986). *Zooplankton community responses to synthetic oil exposure.* In: Cairns, Jr. Ed. Community Toxicity Testing pp. 291-320. ASTM Publication. 350pp.

Hunt, G.L. Jr. (1987). Offshore oil development and seabirds: the present status of knowledge and long term research needs. In: *Long Term*

Environmental Effects of Offshore Oil and Gas Development. D.F. Boesch and N.N. Rabailes [eds.]. London and New York, Elsevier Applied Science Publishers, pp. 539-586.

Ignatiades, L. and Mimicos, N. (1977). Ecological responses of phytoplankton on chronic oil pollution. *Environmental Pollution,* 13, 109-118.

Ivanov, A.Y. and Zatyagalova, V.V. (2008). A GIS approach to mapping oil spills in a marine environment. *International Journal of remote Sensing,* 29(21), 6297-6313.

Jackson, J.B.C., Cubit, J.D., Keller, B.D., Batista, V., Burns, K., Caffey, H.M., et al. (1989). Ecological effects of a major oil spill on Panamanian coastal marine communities. *Science,* 243, 37-44.

Jensen, J.R., Ramsey III, E.W., Holmes, J.M., Michel, J.E., Savitsky, B. and Davis, B.A. (1990). Environmental sensitivity index (ESI) mapping for oil spills using remote sensing and-geographic information system technology. *International Journal of Geographical Information Systems,* 4(2), 181-201.

Jernelov, A. (2010). *The threats from oil spills: now, then and in furute. AMBIO,* 39, 353-366.

Jernelov, A. and Linden, O. (1981). *Ixtoc I: A case study of the worlds largest oil spill. AMBIO,* 10, 299-306.

Johnson, L.L., Arkoosh, M.R., Collier, T.K., Krahn, M.M., Meador, J.P., Myers, M.S., Reichert, W.R. and Stein, J.E. (2001). The effects of polycyclic aromatic hydrocarbons in fish from Puget Sound Washington. In: *The toxicity of fishes.* R. Di Giulio and D. Hinton [eds.]. Taylor & Francis, New York and London.

Jordi, A., Ferrer, M.I., Vizoso, G., Orfila, A. et al. (2006). Scientific management of Mediterranean coastal zone: A hybrid forecasting system for oil spill and search and rescue operations. *Marine Pollution Bulletin,* 53, 361-368.

Junoy, J., Castellanos, C., Vieitez, J.M., de la Huz, M.R. and Lastra, M. (2005). The macroinfauna of the Galicain sandy beaches (NW Spain) affected by the Prestige oil spill. *Marine Pollution Bulletin,* 50, 526-536.

Karydis, M. (1978). Physiological effects of hysrocarbons on phytoplankton. Ph. D. Thesis. University of Wales. 132pp.

Karydis, M. (1980). Uptake of hydrocarbons by the marine diatom *Cyclotella cryptica. Microbial Ecology,* 5, 287-293.

Karydis, M. (1981). The toxicity of crude oil on the marine alga *Skeletonema costatum* (Greville) Cleve in relation to nutrient limitation. *Hydrobiologia,* 85, 137-143.

Karydis, M. (1982a). Marine diatoms affecting the stability of oil-in-water emulsions and hydrocarbon distribution in sea water. *Experientia,* 38, 1185-1186.

Karydis, M. (1982b). Toxicity of a photo-oxidized crude oil on two marine microalgae. *Botanica Marina,* 25, 25-29.

Karydis, M. and Fogg, G.E. (1980). Physiological effects of hydrocarbons on the marine diatom *Cyclotella cryptica. Microbial Ecology,* 6, 281-290.

Keizer, P.D. and D.C. Gordon, Jr. (1973). Detection of trace amounts of oil in sea water by fluorescence spectroscopy. *Journal of the Fisheries Research Board of Canada,* 30, 1039-1046.

Kirby, M.F. and Law, R.J. (2010). Accidental oil spills at sea-Risk, impact, mitigation and the need for co-ordinated post incident monitoring. *Marine Pollution Bulletin,* 60, 797-803.

Kontoes, C.C., Sykioti, O., Paronis, D. and Harisi, A. (2005). Evaluating the performance of the space-borne SAR sensor systems for oil spill detection and sea monitoring over the south-eastern Mediterranean Sea. *International Journal of Remote Sensing,* 26(18), 4029-4044.

Krebs, C.T. and Burns, K.A. (1977). Long term effects of an oil spill on populations of the salt marsh crab *Uca pugnax. Science,* 197, 486-487.

Krishnan, P. (1995). Research report: a geographical information system for oil spills sensitivity mapping in the Shetland Islands (UK). *Ocean and Coastal Management,* 3, 247.

Kulawiak, M., Prospathopoulos, A., Perivoliotis, L., Łuba, M., Kioroglou, S., Stepnowski, A. (2010). Interactive visualization of marine pollution monitoring and forecasting data via a Web-based GIS. *Computers and Geosciences,* 36, 1069–1080.

Lacaze, J.C. (1969). *Effects d' un pollution du Torrey Canyon sur l' algue unicellulaire marine Phaeodactylum tricornutum.* Revue Int. Oceanogr. Med. 13/14, 159-179.

Lacaze, J.C. (1974). Ecotoxicolgy of crude oils and the use of experimental marine ecosystems. *Marine Pollution Bulletin,* 5, 153-156.

Lacaze, J.C. (1983). The importance of controlled ecosystems in studies on marine ecotoxicology. *Physiological and Ecophysiological Impacts of Pollutants on Marine Organisms,* 9, 493-514.

Lacaze, J.C. and Villedon de NeVde, O. (1976). Influence of illumination on phototoxicity of crude oil. *Marine Pollution Bulletin,* 7, 73-76.

Laubier, L., Le Moigne, M., Flammarion, P., Thybaud, E. and Cossa, D. (2004). The monitoring programme of the ecological and ecotoxicological consequences of the "Erika" oil spill. *Aquatic Living Resources,* 17(3), 239-241.

Leahy, J.G. and Colwell, R.R. (1990). Microbial degradation of hydrocarbons in the environment. *Microbiology Review,* 54, 305-315.

Lee, O.A., Olivier, P., Wolt, R., Davis, R.W. and Weltz, F. (2009). Aggregation of Sea Otters (*Enhydra lutris kenyoni*) feeding on fish eggs and kelp in Prime William Sound, Alaska. *American Midland Naturalist,* 161, 401.

Leifer, I, Lehr, W.J., Simecek-Beatty, D., Bradley, E., Clark, R. et al. (2012). State of the art satellite and airborne marine oil spill remote sensing: Application to the BP *Deepwater Horizon* oil spill. Remote Sensing of Environment, 124, 185–209.

Leschine, T.M. (2002). Oil spills and the social amplification and attenuation of risk. *Spill Science and Technology Bulletin,* 7(1-2), 63-73.

Liu, S.K. and Leendertse, J.J. (1986). A three dimensional model of the Gulf of Alaska. Coastal engineering, *American Society of Civil Engineering,* New York.

Martin, S. (2004). *An Introduction to Ocean Remote Sensing*. Cambridge University Press, Cambridge, UK.

Martin, S., Martin, P.H., LeBoeuf, E.J., Daniel, E.B., Dobbins, J.P. and Abkowitz, M.D. (2004). Development of a GIS-based Spill Management Information System. *Journal of Hazardous Materials,* 30, 112(3), 239-52.

McCauley, R.N. (1966). The biological effects of oil pollution in a river. *Limnology and Oceanography,* 11, 475-486.

McKee, E. (1956). Oily substances and their effects on the beneficial uses of water. *California State Water pollution Control Board* (publ. No. 16). Sacramento.

Meador, J.P., Stein, J.E., Reicher, W.L. and Varanasi, U. (1995). Bioaccumulation of polycyclic aromatic hydrocarbons by marine organisms. *Reviews Environmental Contamination and Toxicology,* 143, 79-165.

Mearns, A.J., Reish, D.J., Oshida, P.S. and Ginn, T. (2010). Effects of pollution on marine organisms. *Water Environment Research,* 82(10), 2001-2046.

Miranda, F.P.D., Marmol, A.M.Q., Pedroso, E.C., Beisl, C.H., Welgan, P. and Morales, L.M. (2004). Analysis of Radarsat-1 data for offshore monitoring activities in the Cantarella Complex, Gulf of Mexico using the unsupervised semivariogram textural classifier (USTC). *Canadian Journal of Remote Sensing,* 30, 424-436.

Mommaerts-Billiet, F., (1973). Growth and toxicity tests on the marine nanoplanktonic alga Platymonas tetrathele G.S. West in the presence of crude oil and emulsifiers. *Environmental Pollution,* 4, 261-282.

Monson, D.H., Doak, D.F., Ballachey, B.E., Johnson, A. and Bodkin, J.L. (2000). Long term impacts of the *Exxon Valdez* oil spill on sea otters, assessed through age-dependent mortality patterns. *PNAS,* 97(12), 6562-6567.

Moore, J. (1996). Environmental studies on the Sea Empress oil spill. *Marine Pollution Bulletin,* 32(5), 386-387.

Morrison, M.L. (1988). On sample sizes and reliable information. *Condor,* 90, 275-278.

Nelson-Smith, A. (1972). Oil pollution and marine ecology. *Elek Science,* London, 260pp.

NOAA (1994). *Automated Data Inquiry for Oil Spills Users Manual,* Version 1.1. NOAA Hazardous Materials Response and Assessment Division Report, Seattle, WA.

Nogueira de Andrade, M.M., Szlafsztein, C.F., Souza-Filho, P.W.M., dos Reis Araújo, A. and Tavares Gomes, M.K. (2010). A socioeconomic and natural vulnerability index for oil spills in an Amazonian harbor: A case study using GIS and remote sensing. *Journal of Environmental Management,* 9, 1972-1980.

NRC (1985). Oil in the sea: inputs, fates and effects. *National Research Council.* National Academy Press, Washington, D.C.

NRC (2003). Oil in the sea: inputs, fate and effects. *National Research Council,* National Academy of Sciences, Washington, DC.

O' Brien, G.W., Lawrence, G.M., Williams, A.K., Glenn, K., Barrett, A.G., Lech, M., Edwards, D.S., Cowley, R., Boreham, C.J., Summons, R.E. (2005). Yampi Shelf, Browse Basin, North-West Shelf, Australia: a test-bed for constraining hydrocarbon migration and seepage rates using combinations of 2D and 3D seismic data and multiple, independent Remote Sensing technologies. *Marine and Petroleum Geology,* 22(4), 517-549.

Ogouri, M. and Kanter, R. (1971). Primary productivity in the Santa Barbara Channel. In: *The Biological and Oceanographical Survey of the Santa Barbara oil spill* 1969-1970. Vol. I. (Straughan, D., ed.) 17-42, Allan Hancock Fdn. University of California.

Ohowa, B.O. (2009). Evaluation of the effectiveness of the regulatory regime in the management of oil pollution in Kenya. *Ocean and Coastal Management,* 52, 17-21.

Oil Tanker Spill Statistics (2009). *The International Tanker Owners Pollution Federation.* http:/www.itopf.com/information-services/data-and-statistics/statistics/documents/Statpack2009-FINAL.pdf.

Olita, A, Cucco, A., Simeone, S., Ribotti, A., Fazioli, L., Sorgente, B., Sorgente, R. (2012). Oil spill hazard and risk assessment for the shorelines of a Mediterranean coastal archipelago. *Ocean & Coastal Management,* 57, 44-52.

OSPAR (2007). *Convention of the protection of the marine environment of the North-East Atlantic,* OSPAR Commission, London.

Osullivan, A.J. (1978). Amoco Cadiz oil-spill. *Marine Pollution Bulletin,* 9(5), 123-128.

Parker, C.A., Freegarde, M. and Hatchard, C.G. (1971). The effect of some chemical and biological factors on the degradation of crude oil at sea. In: Water Pollution in Oil (P. Hepple, ed.*) Inst. Of Petroleum,* London, pp. 237-244.

Pavlakis, P., Sieber, A. and Alexandry, S. (1996). On the optimization of spaceborne SAR capacity in oil spill detection and the related hydrodynamic phenomena. *Spill Science and Technology Bulletin,* 3(1/2), 33-40.

Peterson, C.H., Rice, S.D., Short, J.W., Esler, D., Bodkin, J.L., Ballachey, B.E. and Irons, D.B. (2003). Long term ecosystem response to the Exxon Valdez Oil Spill. *Science,* 302, 2082-2086.

Pilpel, N. (1968). The natural fate of oil on the sea. *Endeavour,* 27, 11-13.

Pincinato, F.L., Riedel, P.S. and Milanelli, J.C.C. (2009). Modelling an expert GIS system based on knowledge to evaluate oil spill environmental sensitivity. *Ocean and Coastal Management,* 52, 479–486.

Plutchak, N.B. and Kolpak, R.L. (1981). Numerical simulation of oil spreading on water. *Proc. de La Mechanique des Nappes. Hydrocarbures Assoc.* Amicalle des Ingenieurs, Paris.

Pollani, A., Triantafyllou, G., Petihakis, G., Nittis, K., Dounas, K. and Koutitas, C. (2001). The POSEIDON operational tool for the pre-diction of floating pollutant transport, *Marine Pollution Bulletin,* 43, 270–278.

Prince, R.C. (1993). *Petroleum spill bioremediation in marine environments. Critical Reviews in Microbiology,* 19, 217-239.

Rand, G.M. (Ed.) (1995). *Fundamentals of Aquatic Toxicology: Effects, Environmental Fate and Risk Assessment. 2nd Edition.* Taylor and Francis, London, 1125pp.

Reed, M., Gundlach, E. and Kanay, T. (1989). A coastal zone oil spill model: development and sensitivity studies. *Oil and Chemical Pollution,* 5(6), 411-449.

Reed, M., Johansen, O. Brandvik, P.J, Daling, P., Lewis, A., Fiocco, R., Mackay D. and Prentki, R. (1999). *Oil Spill Modeling towards the Close of the 20th Century: Overview of the State of the Art. Spill Science & Technology Bulletin*, 5(1), 3-16.

Robinson, C.J. and G.L. Cook, 1969. *Low-resolution mass-spectrometric determination of aromatic fractions of petroleum Analytical Chemistry,* 41, 1548–1554.

Romero, A.F., Abessa, D.M.S., Fontes, R.F.C. and Silva, G.H. (2013). *Integrated assessment for establishing an oil environmental vulnerability map: Case study for the Santos Basin region*, Brazil. Marine Pollution Bulletin, 74, 156-164.

Rooker, J.R., Kitchens, L.L., Dance, M.A., Wells, R.J.D., Falterman, B. and Cornic, M. (2013). Spatial, temporal, and habitat-related variation in abundance of pelagic fishes in the Gulf of Mexico: Potential implications of the deepwater horizon oil spill. *Plos One,* 8(10), e76080.

Roth, A.M.F. and Baltz, D.M. (2009). Short term effects of an oil spill on Marsh-Edge fishes and decapods crustaceans. *Estuaries and Coasts,* 32, 565.

Salisbury, J. W., d'Aria, D. M. and Sabins, F.F., Jr. (1993). Thermal infrared remote sensing of crude oil slicks. *Remote Sensing of Environment,* 45, 225–231.

Santos, C.F., Michel, J., Neves, M., Janeiro, J., Andrade, F. and Orbach, M. (2013). Marine spatial planning and oil spill risk analysis: Finding common grounds. *Marine Pollution Bulletin,* 74, 73-81.

Schein, A., Scott, J.A., Mos, L. and Hodson, P.V. (2009). Oil dispersion increases the apparent bioavailability and toxicity of diesel to rainbow trout (*Oncorhynchus mykiss*). *Environmental Toxicology and Chemistry,* 28, 595.

Sebastiao, P. and Soares, C.G. (1995). Modeling the fate of oil spills at sea. *Spill Science & Technology Bulletin,* 2(2/3), 121-131.

Shi, L., Ivanov, A. Yu., He, M. and Zhao, C. (2008). Oil spill mapping in the western part of the East China Sea using synthetic aperture radar imagery. *International Journal of Remote Sensing,* 29(21), 6315-6329.

Smith, J.E. (editor) (1968). *Torrey Canyon pollution and marine life.* Cambridge University Press.

Smith, R.A., Slack, J.R., Wyant, T. and Lanfear, K.J. (1982). *The Oil Spill Risk Analysis model of the U.S. Geological Survey.* Geological Survey, Professional Paper, 1227. United States Government Printing Office, Washington.

Soto, C., Hellebust, J.A., Hutchinson, T.C. and Sawa, T. (1975). Effect of naphthalene and aqueous crude oil extracts on the green flagellate *Chlamydomonas angulosa*. I. Growth. *Canadian Journal of Botany,* 53, 109-117.

Terlizzi, A., Benedetti-Cecchi, L., Bevilacqua, S., Fraschtti, S., Guidetti, P. and Anderson, M.J. (2005). Multivariate and Univariate asymmetrical analyses in environmental impact assessment: a case study of Mediterranean subtidal sessile assemblages. *Marine Ecology Progress Series,* 289, 27-42.

Teruhisa, T., Nakaoka, M., Kawai, H. and Yamamoto, T. (2003). Marine Life Research Group of Takeno, K. Ohwada, "Impacts by Heavy Oil Spill from Nakhodka on Inter - tidal Ecosystem in the Sea of Japan: An Approach to Impact Evaluation with Geographical Information System" , *Marine Pollution Bulletin,* 47, 99-104.

Topouzelis, K., Bernardini, A., Ferraro, G., Meyer-Roux, S., Tarchi, D. (2006). Satellite mapping of oil spills in the Mediterranean Sea. *Fresenius Environmental Bulletin,* 15, 1009-1014.

Topouzelis, K., Karathanassi, V., Pavlakis, P., Rokos, D. (2007). Detection and discrimination between oil spills and look-alike phenomena through Neural Networks. *Journal of Photogrammetry and Remote Sensing,* 62, 264-270.

Torres Palenzuela, J.M., Gonzalez Vilas, L. and Sacau Cuadrado, M. (2006). Use of ASAR images to study the evolution of the *Prestige* oil spill off the Galician coast. *International Jouranl of Remote Sensing,* 27(10), 1931-1950.

Trivero, P., and Biamino, W. (2010). Observing marine pollution with Synthetic Aperture Radar. In P. Impertore, & D. Riccio (Eds.), *Geoscience and remote sensing new achievements* (pp. 397–418). In Tech.

UNEP (1997). Environmental management in oil and gas exploration and production: An overview of issues and management approaches. *Joint E&P Forum/UNEP Technical Publication.*

Varela, M., Bode, A., Lorenzo, J. et al. (2006). The effect of the prestige oil spill on the plankton of the N-NW Spanish coast. *Marine Pollution Bulletin,* 53, 272-286.

Vespe, M., Posada, M., Ferraro, G. and Greidanus, H. (2011). Data fusion of SAR derived features abd ancillary information for automatic oil spill detection. *Fresenius Environmental Bulletin,* 20(1), 36-43.

Wakeman, S.G. (1977). Synchronous fluorescence spectroscopy and its application to indigenous and petroleum-derived hydrocarbons in lacustrine sediments. *Environmental Science and Technology,* 11, 272–276

Westphal, A., and Rowan, M.K. (1970). Some observations on the effects of oil pollution on the Jackass Penguin. *Ostrich* (Suppl. 8:521-526).

Wiens, J.A. and Parker, K.R. (1995). Analyzing the effects of accidental environmental impacts: approaches and assumptions. *Ecological Applications,* 5(4), 1069-1083.

Wirts, K.W. and Liu, X. (2006). Integrating economy, ecology and uncertainty in an oil-spill DSS: The Prestige accident in Spain, 2002. *Estuarine, Coastal and Shelf Science,* 70, 525-532.

Zitko, V. (1975). Aromatic hydrocarbons in aquatic fauna. *Bulletin of Environmental Contamination and Toxicology,* 14, 5.

Zodiatis, G., Lardner, R., Solovyov, D., Panayidou, X. and De Dominicis, M. (2012). Predictions for oil slicks detected from satellite images using MyOcean forecasting data. *Ocean Science,* 8, 1105–1115.

INDEX

#

A

B

C

D

E

F

G

N

O

P

Q

R

S

T

U

V

W

Z